AutoCAD Express NT

Springer
London
Berlin
Heidelberg
New York
Barcelona
Hong Kong
Milan
Paris
Santa Clara
Singapore
Tokyo

Timothy McCarthy

AutoCAD
Express NT

Covering Release 14

 Springer

Dr Timothy McCarthy
Department of Civil and Structural Engineering
UMIST
P.O. Box 88
Manchester
M60 1QD

ISBN 3-540-76155-1 Springer-Verlag Berlin Heidelberg New York

British Library Cataloguing in Publication Data
McCarthy, Timothy J.
    AutoCAD express N.T. : covering Release 14
    1. AutoCAD (Computer file)
    I. Title
    620' . 0042'0285'5369
    ISBN 3540761551

Library of Congress Cataloging-in-Publication Data
McCarthy, Tim, 1959-
    AutoCAD express NT : covering release 14 / Tim McCarthy
        p.   cm.
    Includes index.
    ISBN 3-540-76155-1 (pbk. : alk. paper)
    1. Computer graphics. 2. AutoCAD (computer file) 3. Microsoft
Windows NT.   I. Title
T385.M377651998
604.2'0285'5369--dc21   98-13200

Published by Springer-Verlag London Limited 1999
© Timothy McCarthy 1999
Printed in Great Britain

Typesetting: Camera ready by author
Printed and bound at the Athenæum Press Ltd, Gateshead, Tyne and Wear, UK
69/3830-543210 Printed on acid-free paper

For Fiona and Seán

ACKNOWLEDGEMENTS

I would like to express my deep gratitude to all those who helped me with this project. I am indebted to my colleagues in the Department of Civil and Structural Engineering at UMIST and in the Department of Civil Engineering at Stanford University. Smail Derouiche and Kirti Ruikar helped enormously with testing the exercises. Thanks also to my students from whom I have learned so much and to the Autodesk team in Guildford, UK and San Rafael, California.

My special thanks to Grace, Fiona and Seán for everything.

September 1998   Tim McCarthy

Trademark acknowledgements

CONTENTS

Chapter 1 INTRODUCTION

What is AutoCAD?

AutoCAD is the world's most popular computer-aided drafting package for the personal computer. It is a fully functional 2D and 3D CAD program. Full 3D wire frame representation was incorporated in the program with the launch of Release 10 way back in 1988. These capabilities were enhanced with Releases 11 and 12. R12 saw the first Windows version. Further enhancements, such as the, much praised, ACIS solid modeller and Windows 95 and NT compatibility came with R13 and the current R14. Its popularity has made AutoCAD the de-facto industry standard for PC-CAD with a host of other program developers providing application software conforming to the AutoCAD format.

Poor old Release 13 did not meet with the same great enthusiasm of CAD professionals as earlier releases. While the funcionality had improved it made heavy demands on the hardware. R13 for Windows was actually slower than R12 for DOS for many tasks. Release 14 is the upgrade CAD users have all been waiting for. It firmly marks the emergence of AutoCAD as the major CAD vendor for Windows 95 and Windows NT systems.

New users should be pleased with what is on offer. By fully utilizing the flexibility of the Windows environment and by conforming to the Microsoft user interface standard, AutoCAD R14 is the easiest version to learn yet. Users upgrading from R12 or R13 will notice some welcome changes to the interface that are guaranteed to improve productivity.

As a fully functional drafting program, AutoCAD can achieve anything that can be drawn on a drawing board. The main benefits of CAD come more from being able to edit and exchange drawing information rapidly rather than simply replacing the drawing board. Starting to use AutoCAD is a difficult step as it requires a certain amount of new skill development. Once you have made the commitment to learn how to use the program and implement it in your everyday work the benefits will soon accrue. You will quickly discover that there are many things that you can do with AutoCAD that you could never do with a drawing board.

With AutoCAD your drawings become more than just lines on a sheet of paper. The AutoCAD drawing is a database of information. Some of this is indeed graphic information, but AutoCAD knows the length of every line on the drawing. It knows what symbols and parts have been included on the drawing and it can output this information to design programs or spreadsheets for bill of materials and cost analysis.

The aims of the AutoCAD Express

The main aim of this book is to introduce AutoCAD users to effective CAD drawing techniques. This is done through structured exercises that demonstrate the AutoCAD drafting principles clearly. The commands are dealt with in this context as tools to make the job easier. It is also hoped that you will have fun doing these exercises and creating some of the pretty pictures.

The AutoCAD Express is suitable for new users as it covers the program from the very basics right through to advanced techniques. Occasional users will find it a useful and quick refresher, while even seasoned users will discover novel aspects to old commands. Not only are the commands fully described but AutoCAD *drawing techniques* are explored with many examples.

This edition of the book has been written for AutoCAD Release 14 for Windows NT and Windows 95. It covers all the important aspects of the version with full descriptions of the 3D functions and dynamic viewing. The operating system used is Microsoft Windows 95 and NT Version 4. Where new Windows functions are explained, comparisons with older releases of AutoCAD are given. This will help users who are upgrading to R14, particularly those moving from DOS versions to Windows.

Because it is so flexible, AutoCAD can seem unwieldy to the new user. The exercises in this book follow each other logically along a well structured learning curve. Each chapter represents a stage along this curve, and at each stage the user can pause to consolidate the skills obtained, or proceed to the next stage. To overcome the sheer size of the AutoCAD program and the number of facilities available, the user is directed through the most appropriate path to complete the example drawing. You will never be overwhelmed by lengthy descriptions of abstract concepts and myriad command parameters. Rather you will learn things when you need to know them. By the end of the book there will be little left about AutoCAD that you will still need to know.

The Express route through AutoCAD

The AutoCAD Express is designed as a tutorial guide to the varied facets of the world's most popular computer aided drafting package. The emphasis is on *doing* the various commands and *achieving* results. Chapters 2 to 8 each present instructive drawing exercises which call on AutoCAD's drafting facilities in a logical order. Each chapter has a broad theme with useful asides included where appropriate. Each new command and facility is introduced in the context of solving a particular drafting problem.

Chapter 2 provides a quick introduction to the essentials of producing a drawing file. AutoCAD's brilliant on-line help facility is introduced here too. Chapter 3 gives a complete description of line drawing in AutoCAD with detailed examples of the User–AutoCAD interface. In Chapter 4 the Express takes to the skies to introduce the bulk of AutoCAD's drawing commands. Your first encounter with the program's editing facilities also happens in this chapter. In Chapter 5 the AutoCAD Express lands in Paris, France to explore more advanced editing features and construct A.Gustav Eiffel's famous tower. It's back to the steamy kitchen for Chapter 6 where you will learn how to make and manipulate AutoCAD blocks and create a library of symbols. These symbols are used to help AutoCAD Express Kitchens Inc. to quickly design new fitted kitchens with automatic bills of materials. Their competitors must be worried! Chapter 7 covers automatic dimensioning and a few other high-level commands. The world tour continues in Chapter 8 from the unlikely start back in the kitchen. This particular leg of the journey covers isometric projection and a 2.5D view of the Big Apple before visiting the pyramids of Giza in glorious 3D colour. All the major new facilities introduced for 3D drafting are described with relevant examples covering 3D drawing and visualisation.

Chapter 9 deals with aspects of drawing layout and getting hard copies of your files. This includes a brief description of the printing in Windows and communicating between AutoCAD and other Windows applications. Finally, there is an appendices covering some technical aspects of running and configuring AutoCAD. This appendix also tells you how to customize your own toolbars.

Conventions used in the AutoCAD Express

The style of presentation is fairly simple. There are no distracting icons or hieroglyphics. Plain English is used throughout and where jargon cannot be avoided it is clearly explained. There are a few computerese phrases used in the text which have helped me in writing the book and, I hope, will help you in reading it. Here they are:

RETURN or ENTER?

These are two words that mean the same thing. AutoCAD will frequently tell you that you must "Press RETURN to continue". Now, most keyboards don't have a key called "RETURN" but do have one with "ENTER" or one with an arrow, ↩. All three mean to "enter" the line by pressing the key and "return" the cursor to the left margin. In this book the symbol <ENTER> has been used to signify this. You will also find references to <SPACE> in the book which mean "hit the space bar", gently!

Presentation of user–AutoCAD dialog

What you have to type is shown in bold text. The AutoCAD prompts are shown in normal text. Some points are referenced in diagrams and in the dialog. These references are presented in the text in brackets to the right of dialog. For example:

Command: **LINE** <ENTER>

From point: **35,40** <ENTER>   (V)

This means that AutoCAD will display the word "Command:" and you have to type the word "LINE" followed by pressing the ENTER key. AutoCAD will reply with the prompt "From point:" to which you reply by typing the two numbers separated by a comma and pressing the ENTER key. If you hate typing then you can always use the mouse to pick the point. The "(V)" indicates that this corresponds to the point marked with a "V" on a nearby diagram. You should not type the "(V)".

Most of AutoCAD's commands can be accessed by clicking the appropriate button on one of the toolbars. This will save you having to type the command. When a command button is available as an alternative to typing the command, then it will be shown on the relevant diagram.

Selecting menu items is shown as "pick **File/Exit**" which means, pick the File menu from the top menu bar. This causes a pull-down menu to appear from which you pick Exit.

Control keys

One of the keys on the PC keyboard has "CTRL" written on it. On some keyboards the word is spelt out in full, "Control". When this key is held down simultaneously with other keys special computer commands are executed. AutoCAD and/or Windows use the control key in conjunction with a number of letters to execute different commands. These are given in the text as, say, "CTRL B" or "^B". This means to press the "Control" key and while holding it down also press the "B" key.

Menus

From time to time Autodesk issues improved screen menus. With each new issue, the display details change. Because of this, there may be small discrepancies between the screen menus displayed in this book and those that appear on your screen. Also, AutoCAD's menu interface is fully customizable by the user and so if you or someone else using your system has changed the look of the screen you may have to hunt around for the appropriate toolbar. The tips given in the book will show you how to find the toolbar. There is also a section in Appendix A to show you how to set up your own customized interface. The menus used in the AutoCAD Express are the default ones shipped with the AutoCAD CD-ROM.

Other conventions

Some of AutoCAD's commands require more care than others. Those commands where errors can give disastrous results are preceded by *"HAZARD WARNING!"*. Less dangerous commands are accompanied by a brief "Warning!". Don't avoid these commands. Just follow the safety procedures given with the warnings.

Most pointing devices have more than one button. The "pick button" can be found by trial, though it is usually the left-hand one. Pressing the right-hand button in Windows 95 or NT usually brings up a context sensitive pop-up menu. If the cursor is in the AutoCAD drawing area of the screen, the right button behaves like pressing <ENTER>.

The final note is not about jargon but is timely advice. Be careful not to confuse 0 (zero) with the letter O (Oh), and 1 (one) with l (lower case L).

The Windows environment

Windows' Desktop

If you are not already familiar with using MS Windows then a brief introduction should help. Experienced windowers can leap straight to Chapter 2. What follows is a bare bones tour of Windows and is not meant as a substitute for going through MS Windows' own tutorials.

When you start your computer you are usually confronted by the Windows desktop, similar to one shown in Figure 1.1. The details of what is visible on your system will depend on what applications you have and how many windows are open. It is from the Desktop that all applications are loaded.

To the left of the window shown in Figure 1.1 icons can be found to launch some applications. At the top you will see two important icons. In the corner

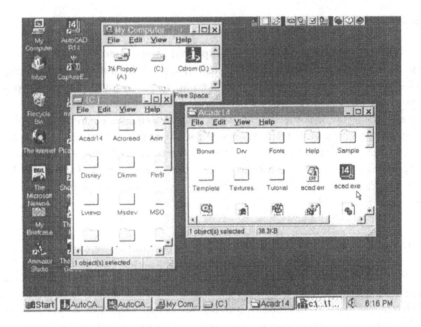

Figure 1.1 Windows Desktop

is the one called "My Computer" and beside it is the AutoCAD icon. Double clicking the R14 icon launches AutoCAD.

If you can't see an AutoCAD icon on your desktop and AutoCAD has already been installed on your system then there are a couple of ways to find it. Double click the My Computer icon to open its window. Move the cursor into the My Computer window and double click on the "C:" to see what is on the C drive. The lower window then shows all the folders on the C drive. If your copy of AutoCAD is on a different drive then double click on that drive's letter. Double clicking on the Acadr14 folder opens a new window in which you will find the icon for acad.exe. The "double click" is two presses of the mouse button in rapid succession.

The "Start" button at the bottom of the screen gives you another method of launching AutoCAD. One click on the Start button pops up a menu as shown in Figure 1.2. Moving the cursor up to "Programs" brings a fly-out menu with all the program groups. You don't need to press the mouse button. Note that the items with a small triangle at their right hand end have another fly-out menu attached. Menu items with no triangle are for executable or readable files.

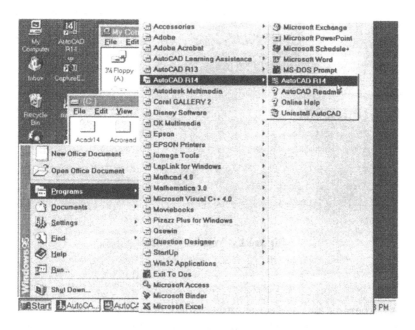

Figure 1.2 Start Menu

If MS Internet Explorer 4 has been installed on your system, the Windows interface will behave differently to that described here. One change is that only one click of the mouse button is required instead of the double click.

Window control

The main features of any window are demonstrated in Figure 1.3. At the top of the window is the title bar. To the left of this is the control icon button. Picking the control button gives the pull-down menu shown in the right-hand diagram. The **Close** allows you to exit the application and close the window. This menu also has options to minimize and maximize the window. Minimize means to reduce the window to a button at the bottom of the screen but without exiting the application. Maximize means to fill the whole screen with this window. On the righthand end of the title bar are three buttons. The one with a single line also minimizes the window. The middle button with the rectangle in it maximizes the window. If a window has already been maximized then this button will show two overlapping tiles which if picked will allow the window to be resized. The final button with the X closes the application and the window.

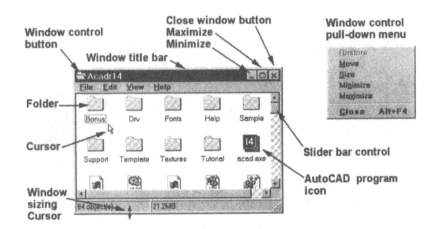

Figure 1.3 Window controls

The programs and files are shown as small pictures or icons in the window. They are opened by moving the cursor onto them and double clicking. You can shift an icon around the window by moving the cursor onto it and pressing the left mouse button. Then, keeping it pressed, drag the icon to a new location. Release the button when you are happy with the new location. If you drag an icon out of the window and drop it in the desktop then it will create a short-cut to that application or file. You can add the AutoCAD R14 icon to your desktop by dragging it from its window onto any clear area of the desktop.

On the window shown in Figure 1.3 some icons are disappearing below the bottom of the window. They can be exposed by dragging the slider bar control down at the right-hand side of the window. Alternatively, the window can be enlarged. If you move the cursor slowly over the frame of the window, it changes into a double headed arrow. This is the window sizing arrow. Now press the button and keeping it pressed drag the frame down, then release. A window can be moved by picking the title bar and dragging it to a new location.

Windows Explorer

Because the windows can be modified or moved with ease the screen can quickly become untidy. Indeed is it possible to "lose" icons and/or files. This is where the **Windows Explorer** comes in, Figure 1.4. This application which is sim-ilar to the old Windows 3.11 *File Manager* is activated in one of two ways. Either pick Explorer from the Start/Programs menu or select File/Explore from the "My Computer" window.

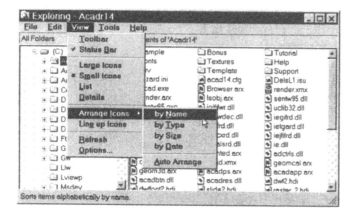

Figure 1.4 Windows Explorer

Windows Explorer allows you to navigate the folders on your computer's disks. If the icon you are looking for is not visible where it should be, you can select **Arrange Icons** from the **View** menu as shown in Figure 1.4. If the file is still missing you can ask Explorer to search the disk for it by selecting **Tools** followed by **Find**.

Windows hot-keys

In windows such as Figure 1.4 the menu commands have one letter underlined. The signifigance of this is that if you hold down the **Alt** key and simultaneously press the underlined letter then that option will be selected.

Here are a few more useful hot keys:

When more than one application has been loaded you can toggle from one to another by ALT and the TAB key. You can also move between them by selecting the appropriate buttons at the bottom of the desktop screen. CTRL and the ESC key gives you the START menu. ESC by itself usually cancels the current operation. The F1 key invokes the HELP facility.

Enjoy the book and soon you will be enjoying the benefits of productive AutoCAD drafting!

Summary

This chapter has given an overview of AutoCAD and the Windows environment.

You should now be able to:

Locate the AutoCAD icon.
Understand the notation used in this book.
Manipulate windows.
Maximize and minimize windows.
Re-size and close windows.
Use Windows Explorer.

Chapter 2 STARTING AUTOCAD

Preparation

This chapter assumes that either Windows 95 or NT and AutoCAD Release 14 have been installed on the computer and are ready to be used. If this is not the case you will need to consult the AutoCAD Installation Guide which comes with your software.

Clear your work area so as to give comfortable access to the computer, keyboard and mouse. An area of about 200mm by 200mm Should be adequate for the mouse and mouse mat. Switch the computer on and wait for it to go through its automatic self test and "booting" procedures and for Windows to start up. This will take approximately 30 seconds. You are now ready to start computer-aided drafting.

The AutoCAD icon

When Windows has loaded you will see a desktop on your monitor much like the one in Figure 1.1. You should be able to see the AutoCAD icon. If so move the cursor arrow tip to the icon and click twice quickly with the left-hand button. After displaying a brief message the AutoCAD window should resemble that shown in Figure 2.1. If you get an AutoCAD window that looks drastically different from Figure 2.1 then check the "Menu preferences" section later in this chapter.

If you cannot see the AutoCAD icon when you start up Windows don't worry. Click on the Start button at the bottom left of the screen. Then navigate with the mouse to "Programs" and "AutoCAD R14" as shown in Figure 1.2.

Creating a drawing

The drawing window

From the Start Up dialog box in Figure 2.1 make sure that Metric units are selected as the default setting and pick the middle box on the left which is

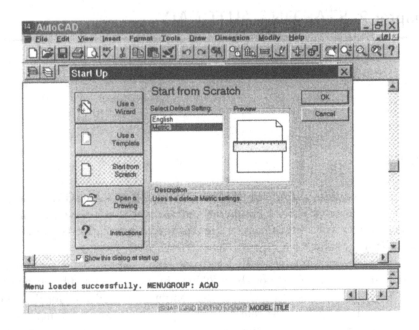

Figure 2.1 Starting AutoCAD

marked "Start from Scratch". Then pick the "Ok" button. Your screen should resemble that shown in Figure 2.2.

Starting at the top of the AutoCAD window you first see the title bar which initially contains the words "AutoCAD - [Drawing.dwg]". This, "Drawing.dwg" will change to the actual drawing name as soon as one is defined. If the word "AutoCAD" appears by itself and the "Drawing.dwg" is on the title bar of another window, here's the score. In AutoCAD you actually have two windows – one for the application and one for the drawing file. You can maximize the drawing window by picking the middle button (the one with the single rectangle) at the right hand end of its title bar.

At the left of the title bar is the AutoCAD Window Control menu icon. This allows you to close the window and the application as described in Chapter 1. At the other end of the title bar are the minimize (horizontal line) and resize (two overlapping rectangles) buttons. This window shown has been maximized. When using AutoCAD it is a good idea to use as much of the screen as possible for the drawing, i.e. maximize the drawing window within AutoCAD and also maximize the application's window.

Immediately below the title bar is the menu bar. There are ten pull-down menus from File through to Help. These contain most of the AutoCAD commands and are described in detail throughout the book. At the righthand

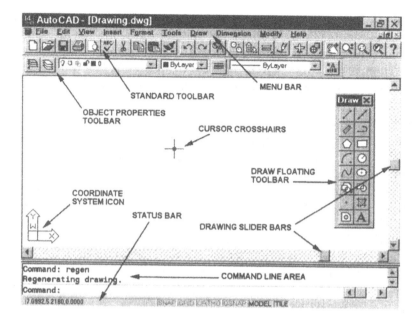

Figure 2.2 AutoCAD drawing window

end of the menu bar you will see the minimize, resize and close buttons for the drawing window.

Under the pull-down menus you will find the AutoCAD Standard "toolbar". This line of icons contains a number of buttons which provide quick access to certain commands. If you move the cursor slowly over the buttons the name of the appropriate command will appear in a "tooltip" pop-up. Clicking the button once activates the command. Below the "Standard" toolbar is the "Object Properties" toolbar. This allows rapid access to different linetypes and colors for objects. If you press the righthand mouse button by mistake when over a toolbar, you will be prompted to edit its properties. As you don't want to do that quite yet, use the Esc key to abort the action.

The bulk of the AutoCAD window is taken up with the active drawing area. This extends from the Object Properties toolbar down to the command line area at the bottom. In the drawing area, the cursor is shown as a pair of cross hairs intersecting at the cursor location. The way the cursor is displayed changes with the context of the operation. When the cursor is moved into the slider, tool bar or pull-down menu area it changes into the Windows arrow for clicking items. Along the bottom of the drawing area and down the right hand side you will see slider bars which are useful for panning around the drawing.

In the drawing area in Figure 2.2 you will also find the Draw toolbar. This is a floating button menu. Each of the icon buttons represents an AutoCAD command. Your tool bar may be displayed differently from Figure 2.2. It may not even appear on your screen right now. Don't worry, we will see how to select toolbars for display in a little while in Figure 2.3 on page 15.

The two little arrows at the lower left of the screen drawing area are AutoCAD's coordinate system icon or symbol. They point to the directions of the coordinate axes. In this case X is the horizontal axis and Y is the vertical. Note that the cursor location in the drawing is shown in a box at the bottom left of the AutoCAD window. In this figure the cursor is 7.0992 units along the X axis and 5.2180 up the Y axis. The third number 0.0000 indicates that, in 3D space, the cursor is zero along the Z axis which is perpendicular to the screen. The "W" indicates that the WORLD or global coordinate system is active. Thie significance of this will be dealt with later, in Chapter 8. The coordinate system icon is on the screen as a reminder to the user, but it is not part of the actual drawing and so does not appear on plots.

The command prompt area is located below the drawing area. This is where commands that you type appear along with the appropriate responses and prompts from AutoCAD. By default, three lines are displayed though the little bitty up/down arrows at the right hand end of the prompt area allow you to scroll to see the last 200 lines.

To the left of the cursor location readout are a number of buttons, "SNAP, GRID" etc. This gives the status of various settings which will be described when appropriate.

AutoCAD's menus

It is possible to type all of AutoCAD's commands using the keyboard. What you type appears on the command line and is executed when you hit <ENTER>. However, to save time and typing errors and also to remind the user of the commands available AutoCAD has a system of menus and submenus. These provide convenient groups of similar commands. The tool buttons provide speedier access to the most frequently used commands. However, some of the icons on the buttons have pretty obscure representations of the commands.

Depending on which version of AutoCAD you are using there may be some slight differences between the menus displayed in this book and those that appear on your screen. A number of vendors supply AutoCAD with customized menus. The menus in *AutoCAD Express* are the standard files with no extra frills.

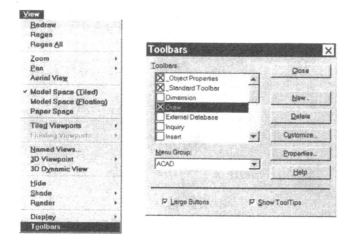

Figure 2.3 Toolbars dialog box

Selecting toolbars

If your AutoCAD window is missing some of the features mentioned above then here's how to go about setting up your menu environment. The toolbars that are displayed are controlled by selecting **View/Toolbars** from the pull-down menu. This is done by moving the cursor into the menu bar at the top of the screen, pressing the mouse button when **View** is highlighted and then selecting the **Toolbars...** at the bottom. Note that pull-down menu items that have 3 dots after them cause dialog boxes to appear. Figure 2.3 shows the setup used for this chapter. The boxes for _Object Properties and _Standard Toolbar should always be checked. Here we also check the **Draw** toolbar. When the toolbars have been selected, click the **Close** button.

A note for experienced "AutoCAD for Dos" Users. If you are used to using AutoCAD's screen menus at the right-hand side of the graphics area then it is still possible to use them. From the **Tools/Preferences...** pull-down menu you get the Preferences dialog box. Clicking the **Display** tab and checking the "Display AutoCAD screen menu" so that an X appears in its box activates the old fashioned menu system. This is not really recommended for new users and is likely to disappear in future AutoCAD releases.

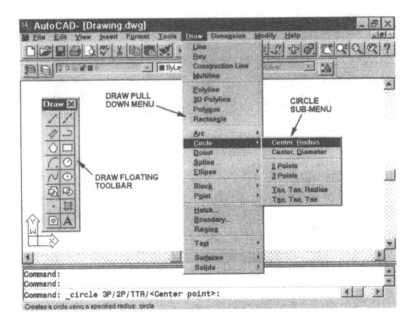

Figure 2.4 Circle sub-menu

How about a bit of doodling?

The menu bar at the top of the window consists of a cascading structure of menus and commands (Figure 2.4). At the end of each branch is a command. As an example, move the cursor so that the **Draw** gets highlighted and press the pick button. The Draw pull-down menu appears. This contains a list of sub-menus and commands. The horizontal lines are purely to visually separate the different groups of commands. Sub-menus are indicated by the triangle symbol pointing right. Menu items with no triangle are executable commands.

Moving the cursor to **Circle** and picking causes the circle sub-menu to cascade as shown in Figure 2.4. Now pick **Center, Radius** and draw a circle. The sub-menu will disappear and at the bottom of the window the command prompt line will display

Command: _circle 3P/2P/TTR/<Center point>:

and will wait for you to input the coordinates of the center. Move the cursor into the drawing area and press the pick button. This is taken as the center point and the prompt changes to

Command: _circle 3P/2P/TTR/<Center point>: Diameter/<Radius>:

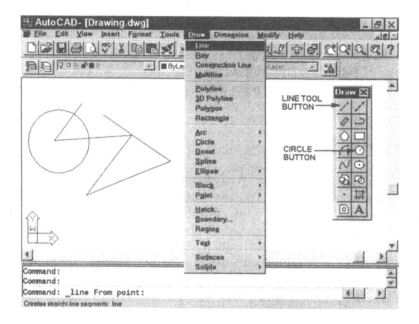

Figure 2.5 Doodles

waiting for the value of the radius to be input. As you drag the cursor the radius appears and the circle grows and shrinks with the cursor position. When you are happy with the size of the circle press the pick button.

To draw some lines pick **Draw/Line**. When you get the "From point:" prompt, pick a location on the drawing area.

Command: _line From point:

The prompt changes to "To point:". Pick as many points as you want to create a series of connected line segments (Figure 2.5). To exit from the LINE command simply press **<ENTER>** or the right mouse button (as long as the cursor is in the drawing area of the screen).

This has been a quick squirrel hop along one part of the menu tree. Tasty morsels are also to be had along the other branches. As you use AutoCAD you will become familiar with the menus and the most useful commands. If you get lost on the menu tree, pressing **ESC** will return you to the graphics screen. Remember, commands can also be typed at the keyboard or picked from the floating tool boxes.

Note that the commands which were executed by AutoCAD following button and menu picks were preceded by the underscore character, eg _circle.

These are the basic command names. AutoCAD uses an alias or dictionary file to map the local language of the user to the basic underlying command name. thus in English you type "CIRCLE" and AutoCAD executes _circle; in French, you type "CERCLE" and AutoCAD still executes _circle. You can therefore ignore the underscores as the more usual name for the command works too.

The tool boxes

Now that AutoCAD is a true Windows application it has a huge array of button menus or toolboxes. We have already seen how to make the Draw toolbox appear (by selecting **View/Toolbars**, Figure 2.3. There are a wide number of toolbars available and these give super fast access to AutoCAD's commands. They are all floating toolbars, which means that they can be moved around the window to convenient locations. Figure 2.6 shows various ways the Draw toolbar can be displayed. This particular image is a composite one. Only one version of the draw toolbar can be shown at a time. This contains 16 of the most frequently used drawing commands. The toolbar can be moved around like any window, i.e. by picking its title bar and dragging it around. It can be resized by moving the cursor slowly close to the edge. When the resizing arrows appear, the shape of the toolbar can be altered. If you drag the toolbar to any of the edges of the drawing window then it will "dock" itself. To undock a toolbar, pick the border of it and drag away from the edge of the window. You can avoid the automatic docking by pressing the CTRL key while dragging the toolbar.

The icons are a bit obscure to start with but if you move the cursor slowly over them the name of the command appears in a tooltip pop-up. The LINE command can be executed by picking the button on the top left of the floating toolbar (two dots joined by a straight line). Note that the status line at the bottom of Figure 2.6 changes to give a brief description of the command too.

Setting up the drawing environment

Now that you know a bit about communicating with AutoCAD, why not execute some commands and embark on some controlled CAD? The emphasis here is on your being in control and not accepting any old rubbish the computer might tempt you with in the name of "convenient defaults". The default drawing environment contains all the settings somebody somewhere found suitable for his or her own application. You can be sure that they were not meant for you. Even if it is acceptable for one drawing the environment may need changes for the next.

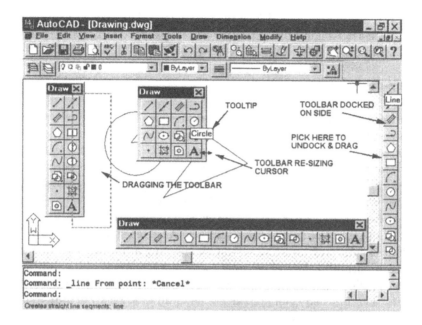

Figure 2.6 Draw toolbar (composite image)

For this section you will need a clear drawing, so if you have already drawn some doodles, follow the procedure below to discard the current drawing. If your drawing is empty, then skip to the "Drawing size" paragraph.

Pick **File** from the menu bar at the top of the window followed by **New**. The dialog box in Figure 2.7 will appear to warn you that the drawing has not been saved. You now have the option to save or discard the changes to the drawing file or to cancel the operation and return to the drawing editor. In this case pick **No** to dump the unwanted doodles.

This will discard the current drawing and then display the "Create New Drawing" dialog box. Select Metric units and pick the "Start from Scratch"

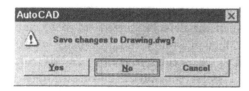

Figure 2.7 Dumping an unwanted drawing

option just like Figure 2.1. Click on the **OK** button. This starts a new drawing, again called Drawing.dwg.

Drawing size

All AutoCAD drawings are made to *full scale*. One of the principal reasons for this is so that when you use AutoCAD's automatic dimensioning it will give the correct lengths rather than scaled ones. Thus the drawing size will depend on the size of the items being drawn. It will also depend on the working units. For example, to draw the architectural layout of an office measuring 24m by 15m you would choose a size big enough to contain the whole floor, plus a bit to spare for extra views. A good drawing size in this case might be 29700 by 21000. (Note that this is the same ratio as an A4 sheet and a base unit of 1mm is assumed.) Although it is not essential, choosing a drawing size that bears some relationship to the size of paper facilitates the layout of the drawing for subsequent plotting. For the new drawing you will require it to be 65 units by 45 units. The LIMITS command allows us to do this. Pick **Format** from the menu bar and then **Drawing Limits**. This executes the LIMITS command and the command line area should echo the following. What you have to type is in bold text.

> Command: 'limits
> Reset model space limits:
> ON/OFF<Lower left corner> <0.00,0.00>: <ENTER>
> Upper right corner <420.0000,297.0000>: **65,45** <ENTER>
> Command: **LIMITS** <ENTER>
> Reset model space limits:
> ON/OFF<Lower left corner> <0.00,0.00>: **ON** <ENTER>
> Command:

Pressing <ENTER> in response to the lower left prompt means that you accept the default setting of positioning that corner at the WORLD origin, 0,0. AutoCAD always displays the default values between angular brackets, < >, and any time you wish to accept this value simply press <ENTER>. If you wish to override the default then key in the desired values as in the "Upper right" prompt above. If the defaults offered by your computer are different from 420.0000,297.0000, don't worry. Simply replace whatever the values are by **65,45** <ENTER>. The second execution of LIMITS is to turn the limit checking facility on. This prevents anything being drawn outside the limits by mistake. This is especially important when using AutoCAD for Windows since the shape and size of the drawing window usually does not match the actual limit proportions.

At this point if you move the cursor around the drawing area you will notice that the coordinate read-out in the status bar is giving similar values as before. So, even though you have changed the drawing size the display shows the old size. To display the whole of the new drawing size type:

Command: **ZOOM** **<ENTER>**
All/Center/.../Scale(X/XP)/<Realtime>: **A**

The response to the type of zoom required can be truncated to whatever Auto-CAD displays in CAPITAL letters, in this case "A" for "All". This command can be found by picking **View** from the menu bar, then **Zoom** followed by **All**. It's a toss up whether it is quicker to type or to pick from the menus. This command works like a zoom lens in a camera allowing magnification and demagnification of the image. ZOOM is fully described in Chapter 3. Now, if you move the cursor around, the coordinates will reflect the current drawing size.

While it is important to consider the drawing size before you embark on an AutoCAD drafting session, it is not essential. You can alter the drawing size at any time during editing, but it's more efficient to get it right first time!

If the cursor now appears a bit jumpy when moved about the drawing area you can adjust the snap increment. The SNAP value, when enabled, limits cursor movements to discrete steps. If the steps are too large the cursor will seem jumpy. If it is too small or switched off then it can be difficult to accurately pick points in the drawing. The following short command sequence sets a snap value of 1 unit which is about right for the subsequent exercise. In this instance the command is typed at the keyboard. Further examples of SNAP are given in Chapter 3.

Command: **SNAP** **<ENTER>**
Snap spacing or ON/OFF/Aspect/Rotate/Style <>: **1**

The next choice you have for the drawing environment is the system of units to be used. Standard default settings use decimal units and decimal angles. Generally, the base unit is the inch but the exact meaning of 1 unit is up to the person creating the drawing. In a building plan, inches might be fine while for a map then miles might be more appropriate. Obviously, if metric units are desired then that too is possible. Here we are going to select Decimal units with a precision of 2 places after the decimal point. Decimal degrees with a precision of 1 place after the point will be chosen for angles. To set the units, pick **Format** from the menu bar and then **Units...** The dialog box shown in Figure 2.8 should appear. Pick **Decimal** in the Units column and **Decimal Degrees** for angles.

To adjust the precision of the coordinate readout click on the Windows pull-down icon (down pointing arrow) to the right of the box below the word

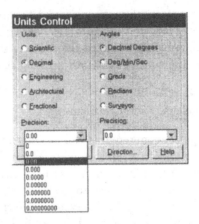

Figure 2.8 Units control settings

"Precision". This brings the pull-down menu with varying numbers of zeros after the decimal point. Use the scroll arrows on the right to move up or down. When **0.00** is highlighted, pick it with the cursor. Repeat this procedure for the an angular precision of **0.0**.

Finally, with angles you can choose which direction is represented by 0 degrees. Pick **Direction** from the bottom of the dialog box to make sure that East is 0 degrees. Click **OK** once to close the Direction Control box and once more to close the Units Control box. This procedure should ensure that the examples in this book match the displays on your screen.

There are many other drawing "environment" settings that can be specified but for the time being we will leave it at Limits and Units. Other settings will be introduced when they become useful.

Drawing lines

Having done all that work on setting up the drawing environment, you are in a position to do some controlled drafting. The workhorse of any drawing is the humble line. Indeed, ultimately, every object can be reduced to a series of straight lines (curves consist of a very large number of tiny straight lines).

The AutoCAD LINE command allows you to construct any number of independent line segments. To draw a square 15 x 15 pick **Draw/Line** in that order from the menu bar and type the coordinates given below. In response to the final "To point:" prompt type the word "close".

Command: LINE

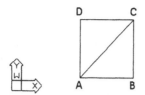

Figure 2.9 Drawing a square

From point: **5,5** **<ENTER>** (A)
To point: **20,5** **<ENTER>** (B)
To point: **20,20** **<ENTER>** (C)
To point: **5,20** **<ENTER>** (D)
To point: **CLOSE** **<ENTER>**

AutoCAD's first response to the LINE command is to ask for a start point and then the end point. The program assumes that you want to continue drawing lines until you tell it otherwise. By typing "**CLOSE <ENTER>**" you cause AutoCAD to join the end point of the last line segment to the very first point input. Now draw a line along one of the diagonals of the square from A to C in Figure 2.9. These letters won't appear in your drawing, they are just used to clarify this text.

Command: **LINE <ENTER>**
From point: **5,5** **<ENTER>** (A)
To point:

If you move the cursor around the drawing area the line "rubber bands" and extends from the point 5,5 to wherever the cursor is located. As the line rubber-bands the coordinate display on the tool bar may jump to polar notation ie distance and angle. If this happens press ˆ**D** (CTRL+D) to toggle it back to x,y coordinates.

You can now pick the point at the other end of the diagonal or type the coordinates.

To point: **20,20** **<ENTER>** (C)
To point: **<ENTER>**

As mentioned above, AutoCAD keeps replying with the prompt "To point:". When you have finished inputting lines you can exit the LINE command by pressing **<ENTER>** without giving any point. Alternatively, you can hit the space bar on the keyboard instead of the ENTER key. In most AutoCAD com-

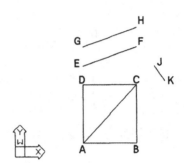

Figure 2.10 Adding some lines

mands it is possible to interchange use of the space bar and the ENTER key.
As the former is much larger it is much easier to locate (or harder to miss!). A
third way to get out of the LINE command (and indeed any other AutoCAD
command) is to CANCEL it. Press **Esc**, the escape key on the keyboard to
cancel the current operation and return to the "Command:" prompt. Remember **Esc** as it can get you out of trouble when things go wrong. CANCEL
appears in dialog boxes, ready to come to the rescue.

What if you want to draw two lines which are not connected? Use
Draw/Line twice. AutoCAD gives some assistance by using the same LINE
command as used above but adds an <ENTER> after the second point has
been picked. To draw the two parallel lines, EF and GH, in Figure 2.10 pick
Draw from the menu bar and then **Line**. Then pick the point 5,25 and pick
or type the second point 20,30. Press **<ENTER>** to exit LINE and return to
the command prompt.

Command: LINE From point: **pick 5,25**   (E)

To point: **20,30 <SPACE BAR>**   (F)

To point: **Pick Right button on mouse** or press <ENTER>

Now pick the **Draw/Line** sequence once more.

Command: LINE From point: **5,30 <ENTER>**   (G)

To point: **20,35 <ENTER>**   (H)

To point: <ENTER>

Command:

AutoCAD's command memory

AutoCAD remembers the last command executed and it offers that command
again if you press <ENTER> or <SPACE BAR> at the "Command:"
prompt without typing anything else. The right-hand mouse button also has
this function when the cursor is in the drawing area of the screen. This is an
obvious time saver when doing repetitious sequences like drawing lots of lines.

To draw the line JK in Figure 2.10 press <ENTER> and pick the points
25,25 and 28,21 and press <ENTER> to get back to the Command: prompt.

Command: <ENTER>
LINE From point: **25,25**   (J)
To point: **28,21**   (K)
To point: <ENTER>
Command:

We have already covered the "close" option. It makes a closed polygon of the
lines by joining the most recently input point to the first point (ie. the "From
point:") of the sequence. Close, which can be abbreviated to "c", will only
work if at least two lines have been drawn in the *current* LINE operation. The
"current LINE operation" means all the inputs from ONE issuing of "LINE"
at the Command prompt.

Another sub-command that only works within the current LINE operation
is the "undo" option. If, while drawing a sequence of connected line segments,
an incorrect "To point:" is picked, you can immediately undo the last pick and
re-input a new point. To illustrate, pick the **Draw/Line** from the menu bar
and then the points P, Q, R and S' (see Figure 2.11). Without leaving the LINE
command type **undo** at the "To point:" prompt. The point S' disappears and
you are prompted for a new "To point:" and can pick the points S and T.

Command: **LINE** <ENTER> From point: **35,5** <ENTER>   (P)
To point: **35,10** <ENTER>   (Q)
To point: **40,15** <ENTER>   (R)
To point: **45,15** <ENTER>   (S')
To point: **undo** <ENTER>
To point: **40,20** <ENTER>   (S)
To point: **35,20** <ENTER>   (T)
To point: <ENTER>

You can undo all the line segments back to the point, P, in this manner but
only if they are all part of the same LINE operation.

There is another UNDO command in AutoCAD which can reverse any
previous command. The one within the LINE command only works for back-

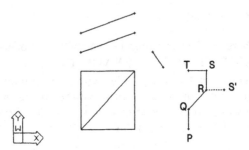

Figure 2.11 The undoing of a line

tracking the line segments. If you type UNDO at the "Command:" prompt it will back-track the previous command or commands. This is described further in Chapter 4.

HAZARD WARNING! Typing UNDO at the Command: prompt can be the undoing of everything. If you undo too much, use the REDO command immediately.

Linetypes and scales and colors

All the lines drawn so far have been standard continuous ones, either white on black or black on a white background. Engineers and architects use many different linetypes to signify different things. Center lines, hidden lines, dashed lines etc are all commonly used in drawings. As you would expect AutoCAD contains definitions of all the important linetypes, including various International Standard Organisation (ISO) definitions. Colors are also used to indicate certain types of information.

AutoCAD gives two ways of assigning linetypes and colors. The method given in this exercise is the "quick and nasty" method. We just tell AutoCAD to change the linetype and color for subsequent objects (lines, circles, text etc). The second method, described in Chapter 3, is to divide the information on the drawing into a series of "layers". For example, the plans of a house might contain a layer of plumbing information, a layer of electrical layout and layers for walls etc. Each layer is then assigned an appropriate default color and linetype. You then have to tell AutoCAD which layer you wish to work on.

Getting back to the quick and nasty method, take a closer look at the toolbar at the top of the drawing area (Figure 2.12). In the current drawing, Drawing.dwg, there is only one layer which is called "0". This name appears in

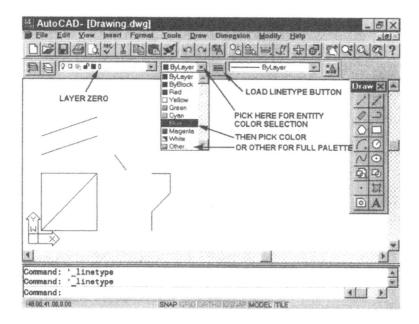

Figure 2.12 Object Color selection

the object properties toolbar along with a series of little icons with a lightbulb and sun. Just to the right of the "Layer zero" information is a square which indicates the current color beside the word "Bylayer". Moving the cursor into the "color box" and clicking also gives a pull-down list of available colors.

The initial color is called "BYLAYER", ie whatever the layer has been set to. The default layer color is white on black (or black on white depending how your screen is configured). Pick **Blue** from the list. If this is not your favorite color try picking **Other...** which brings up the full color palette from which you can pick your favorite shade of blue and pick **OK**. Note that if you select a color from the standard list, then its name appears in the object properties toolbar. The little color swatch beside the name should also change to blue. If you pick one from the full palette then that color number appears.

Note: Everything you draw from now on will appear blue – until you change the color settings. Colors can also be changed by picking **Format/Colors** from the menu bar or the command, COLOR, can be typed.

Just as the basic default for color is black, the default for linetype is "Continuous". To draw dashed and other linetypes you must first load the linetype

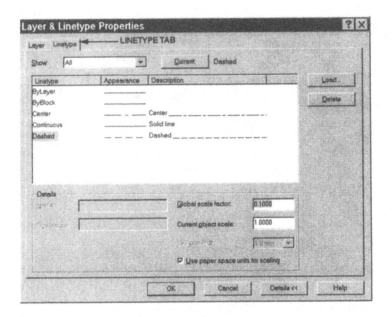

Figure 2.13 Layer & Linetype Properties dialog box

definition into your drawing. Once they are loaded, they can be selected for object creation in a similar way to color selection.

Pick the **Linetype** icon shown in Figure 2.12 from the middle of the Object Properties toolbar. This brings up the Layer & Linetype Properties dialog box shown in Figure 2.13. Make sure that the Linetype tab is at the front. Pick the tab to make sure.

Center and Dashed will not appear on your list, initially. Try picking the **Load** button. This brings another dialog box with a long list of different linetypes (Figure 2.14). Use the scroll bar to move down the list to the linetype, CENTER. Pick, this to highlight it and select the OK button. Repeat the process for the DASHED linetype.

If you don't get a list containing the dashed and center line definitions, check that the file "acad.lin" is being used. You can see the filename at the top of the dialog box in Figure 2.14. If a different filename is given then pick the **File** button and select the correct file.

Your Linetype Properties dialog box should look like the one in Figure 2.13. The final things to do are to make Dashed the current linetype. Pick the word **Dashed** so that it is highlighted and then pick the **Current** button towards the top of the dialog box. Then pick **OK** at the bottom of the dialog box.

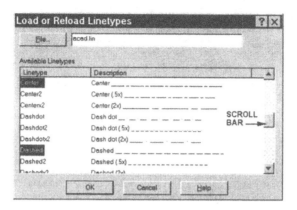

Figure 2.14 Loading Linetypes

Now draw the lines U to V to W to X to close as shown in Figure 2.15.

Command: **LINE <ENTER>**
From point: **60,40 <ENTER>**   (U)
To point: **35,40 <ENTER>**   (V)
To point: **35,30 <ENTER>**   (W)
To point: **60,30 <ENTER>**   (X)
To point: **C <ENTER>**

Did the lines come up dashed? If not, this may be because the dashes are too small or too big. You can control the size of the dashes by altering the Global linetype scale factor. Initially, the scale factor has a value of 1. For the DASHED linetype this means that you have one dash and one space per drawing unit. This would have worked well if we had not altered the drawing limits. Since we reduced the limits by a factor of just less than ten and for consistency with Figure 2.15 set the value to 0.1.

The easiest way to do this is to pick the Linetype button from the Object Properties toolbar as before and bring up the Properties dialog box again (Figure 2.13). In the area marked "Details" there is an input field for the "Global scale factor". Move the cursor into the input field and press the mouse button. Change this value to **0.1** and pick **OK**. The dashes should now be twice as big as before.

Note for AutoCAD 12 users This is exactly the same as setting the LTSCALE value. The LTSCALE command still works but the dialog box gives you an extra option. Because this is a global scale factor it will be applied to all objects in the drawing. If you want to have different dash lengths for individual objects

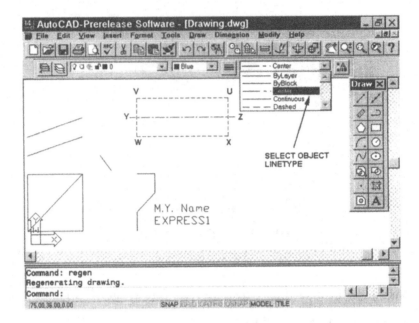

Figure 2.15 Dashed lines

then use the "Current object scale" in the dialog box. The actual dash length will be the product of the global scale and the current object scale. While the LTSCALE command still works as in Release 12, the Linetype command now always brings up the dialog box.

If nothing seems to happen try regenerating the drawing:

Command: **REGEN <ENTER>**

Regenerating drawing.

The REGEN command forces AutoCAD to re-calculate all the lines using this new linetype scale. Depending on your default settings this regeneration may occur automatically after altering the linetype scale. The larger the drawing limits the larger Global scale factor so that the dashes are visible on the screen. The real test for linetype scales is if the dashes are the right size on the final plot. You may need to experiment.

To draw the final centerline on the blue dashed rectangle we need to change the linetype. Because the linetype "Center" has already been loaded in the drawing file it can now be selected from the Linetype pull down menu shown in Figure 2.15. Pick the Object linetype box, which should have the word "Dashed". The pull down menu listing all the loaded linetypes should appear. Now pick **Center**. You could still have used the Linetype dialog box as before, but this method is much quicker. Now draw the line.

    Command: **LINE <ENTER>**
    From point: **33,35 <ENTER>**   (Y)
    To point: **62,35 <ENTER>**   (Z)
    To point: **<ENTER>**

Only linetypes that have been loaded into the current drawing will appear in the pull down list or in the Linetype dialog box. While this quick and nasty method gives flexibility to your drafting, it must be used with caution. Since different types of line usually convey different types of information it makes sense to collect them onto individual layers as described in the next chapter. If at all possible use only one linetype per layer.

WARNING! When we are finished with the special colors and linetypes it is advisable to reset them to the **BYLAYER** option. This can be done from the pull down list of colors and the list of linetypes each of which contain ByLayer as an item. If you don't get into the habit of reverting the settings to ByLayer, you will find that objects will frequently be drawn with the wrong color and/or linetype. This will necessitate extra editing at a later stage.

Saving the drawing

While working in AutoCAD you can periodically save the drawing and any changes you have made to it. The current drawing name is the default, Drawing.dwg. The first time you try to save this file to disk, AutoCAD will prompt you for a name for the file. It is a good idea to choose a different name. The name you give can be any valid Windows 95 or NT filename. AutoCAD adds ".dwg" as the filename extenstion.

To save Drawing.dwg with the new name EXPRESS1 pick **File** from the menu bar followed by **Save....** Alternatively, type **SAVE <ENTER>** at the Command prompt. This displays the "Save Drawing As" dialog box (Figure 2.16). Click **OK** to accept the default drawing name, which in this case is "EXPRESS1".

If you do not get the Save file dialog box then check the value of the FILEDIA variable. If this is set to 0, some dialog boxes are suppressed. It should be set to 1.

    Command: **SAVE <ENTER>**
    Save Drawing As <Drawing.dwg> **EXPRESS1**
    Command: **FILEDIA**

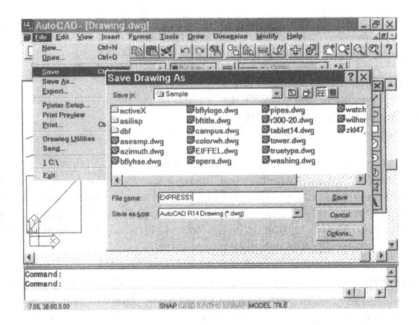

Figure 2.16 Save Drawing As dialog box

New value for FILEDIA <0>: **1**

You should use SAVE approximately every 15 to 20 minutes during a
drawing session. Always do a SAVE before attempting difficult or large tasks
(eg copying a large layout) or before attempting something new. Some commands are categorized in this book as potentially dangerous (HATCH, HIDE,
Print etc) and a SAVE should be done before executing them.

If the file EXPRESS1.DWG already existed on the disk you would get
a warning message from AutoCAD. If you then proceed with the SAVE, the
old EXPRESS1.DWG file will be renamed EXPRESS1.BAK and a new EX-
PRESS1.DWG file made. This backup file will be stored in the same directory
as the original .DWG file.

Autosave setting

There is an autosave option in the AutoCAD Preferences dialog box (see Figure 2.17). This is really useful and is simple to set up. Pick **Tools/Preferences**
from the menu bar. Then pick the **General** tab. Make sure the Automatic save
is ticked and the time between saves set to 20 minutes. The "Create backup

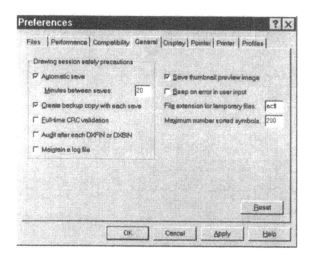

Figure 2.17 Preferences dialog box

copy" should also be ticked. The other stuff is not so important at this stage. Then pick the **OK** button.

Adding some text

Before you leave EXPRESS1 as a completed drawing, you can put your name to it, as shown in Figure 2.15. You can insert your own name in place of "M.Y. Name". Pick **Draw**, then **Text** and **Single line text** from the menu bar. This executes a command called DTEXT. There will be more about text and multiline text in chapter 4.

> Command: **DTEXT** **<ENTER>**
> Justify/Style <Start point>: **40,10** **<ENTER>**
> Height <0.20>: **2** **<ENTER>**
> Rotation angle <0>: **<ENTER>**
> Text: **M.Y. Name** **<ENTER>**
> Text: **EXPRESS1** **<ENTER>**
> Text: **<ENTER>**

Figure 2.18 AutoCAD Help contents

HELP!

That piece of text completes the drawing part of this chapter. Before finishing this editing session it's worth taking a quick look at AutoCAD's HELP facility. This provides information about all of AutoCAD's commands, system variables and all the Windows specific features of the program. It is structured in a similar manner to Help facilities for other Windows applications.

As with Windows the AutoCAD help facility can be activated by pressing the **F1** key. It can also be found at the right-hand end of the menu bar and at the end of the Standard Toolbar (the button with the big ? on it). From the menu bar pick **Help** followed by **AutoCAD Help Topics**. The AutoCAD Help window opens as shown in Figure 2.18. Clicking on any of the book icons opens that topic to give a list of chapters which appear as questionmarks and chapter titles. When help extends over one screen then the scroll bar at the right can be used.

You can find yourself moving down thru the levels of help quite quickly. If you get lost, the "Help Topics" button will bring you back to Figure 2.18 while the "Back" button steps back one screen at a time. Pressing F1 again at the contents screen gives you help about the Help facilities. You can exit HELP and get back to the drawing screen by picking **File** and **Exit** or by picking the close button marked "X" in the top right corner of the Help Window.

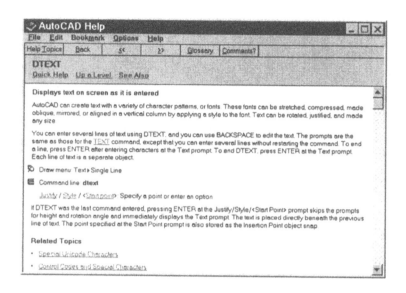

Figure 2.19 DTEXT help

By way of an example pick **Command Reference** from the Contents section shown in Figure 2.18. Then pick **Commands**. This gives a screen-full of new topics in alphabetical order. Now pick the **D** button from the row of letters near the top and scroll down until you find DTEXT. Click on **DTEXT** to find out all about that command (Figure 2.19). Related topics appear as green text with solid underline. Click on these words for information about them. Items with dotted underlined green text lead to pop-up definitions. Click the mouse anywhere to make the definition disappear.

The "Find" tab in Figure 2.18 is a quick way of finding help on a specific command or topic. Clicking the **Find** tab gives another dialog which allows you to input a search string. As you type the search string Help will return a list of topics that can be displayed.

Context sensitive Help

If you know the name of a command but are unsure about how to use it then the context sensitive help is a neat solution. When you are in the middle of an operation, say the line command, press **F1** or pick the help button. The information displayed will relate to the command in progress.

Command: **LINE <ENTER>**
From point: press **F1** key   Or press the Help button

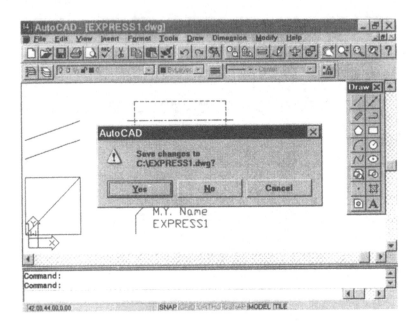

Figure 2.20 Save changes dialog box

This will tell you all about the line command. You can exit help as before and the LINE command will resume.

Spend some time exploring AutoCAD's help features. Of particular note for new users is the section on Tutorials (Figure 2.18) and the "How to" section. The tutorials have demo features which run thru the AutoCAD commands to create a number of drawings. These are a useful resource for the learner.

Finishing up

To finish a drawing editing session you can either start editing another drawing or exit AutoCAD altogether. To switch to a new drawing pick **File/New**, to edit an existing drawing pick **File/Open** or to leave AutoCAD pick **File/Exit**. As we have not yet saved the last changes, an alert dialog box will pop up again as shown in Figure 2.20.

Pick **Yes** and the updated EXPRESS1.dwg will be saved and AutoCAD will shut down, returning you to Windows.

NB: Keep a safe copy of this drawing file as it will be useful for the exercise in Chapter 4.

If you do not want to leave the drawing just yet, pick **Cancel**. If you did not want to keep the new drawing (or changes to an old one) you can leave the editor by picking **No**. You can also do this by issuing the QUIT command. This brings up the same dialog box as before if there are unsaved modifications to the file.

Command: QUIT <ENTER>

This completes your first non-stop journey on the AutoCAD Express. In subsequent chapters it will be assumed that <**ENTER**> is pressed at the end of each command line or that the commands are picked from menus.

A note on file security

Many things can go wrong with computers and disks and occasionally with AutoCAD itself. The words "FATAL ERROR" have an ominous ring and it is very annoying when such an event occurs. The only sane way to approach the inevitable is to prepare well, so that when it happens the consequences are minimized. Don't be depressed by all this doom and gloom. With adequate preparation the impact of a computer crash is cushioned.

It is good practice to have more than one copy of your work. AutoCAD provides facilities for this in two ways. Firstly, if you are going to edit an existing drawing file, then AutoCAD automatically makes a copy of the drawing. If the original was called "EXPRESS1.DWG" then the backup copy is called "EXPRESS1.BAK". This is certainly a very useful security measure but in itself is not fool proof. You can use the Windows Explorer to make further copies onto other disks or onto a DAT or DVD device. With all important computer files it is recommended to have at least two copies on your working disk, and one safe copy on another disk which should be kept in a safe place. Regular backing up of files is *essential.*

Summary

This chapter has introduced many of AutoCAD's facilities and its methodology. In this way the chapter provides the basis for understanding the way all the commands work.

You should now be able to:

Load the AutoCAD program
Create a new drawing file.
Pick commands from pull-down menus and the toolbars.
Move and re-size toolbars.

Set up suitable AutoCAD drawing limits.
Draw lines and squares and circles.
Undo lines.
Use dialog boxes to select files.
Load extra linetypes.
Assign linetypes and colors to objects.
Alter the linetype scale.
Automatically save a drawing file every 20 minutes.
Make extra backup files.
Navigate the AutoCAD Help Facility.
Exit from AutoCAD, either saving or discarding changes.

Chapter 3 CURSOR AND DISPLAY CONTROL

General

Drawing with a mouse or tablet cursor is analogous to drawing freehand with a pencil. In engineering drawing we rely on a host of drawing aids and equipment. T-squares, rulers, setsquares and compass are but a few of the tools necessary to control the positions and sizes of drawn objects.

Similarly, we need electronic protractors and digital dividers when using AutoCAD. This chapter covers some of the useful cursor control facilities within AutoCAD. Using these facilities will make your drawings into exact graphical representations. New types of construction lines were introduced in Releases 13 and 14. You will learn how to set up construction lines of infinite length. These will then be used to reference drawing items to the key construction points for making a 40m replica of Paris's Eiffel Tower. The half tower shown in Figure 3.1 is the goal of the current exercise. The tower will be completed in Chapter 5.

For this chapter you will create a new drawing file, called **EXPRESS2**. The first task is to set up the drawing environment. Using the limits of (0,0) to (65,45) you will then load a number of AutoCAD's linetypes. As this drawing will contain a number of different categories of information you will be introduced to AutoCAD's layering facility.

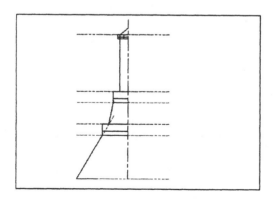

Figure 3.1 The Half full tower

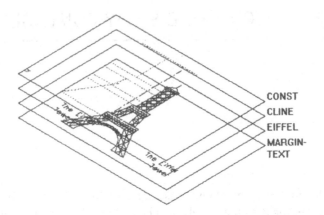

Figure 3.2 Layers as transparent sheets

Start up AutoCAD as described in Chapter 2. Set up the drawing environment using LIMITS and UNITS.

Command: **DDUNITS** or pick **Format/Units...** from the menu bar.

Then pick **Decimal** units and **Decimal Degrees** for the angles. Set the precision for linear units to **0.00** and the angles to **0.0**. Then pick **OK** to set the units. The LIMITS command is also on the **Format** pull-down menu and is called **Drawing Limits**.

Command: **LIMITS**
Reset Model space limits:
ON/OFF/<Lower left corner> <0.00,0.00>:
Upper right corner <420.00,297.00>: **65,45**
Command: **LIMITS**
Reset Model space limits:
ON/OFF/<Lower left corner> <0.00,0.00>: **ON**
Command: **ZOOM**
All/Centre/.../Scale(X/XP)/<Realtime>:**A**

Remember turning the limits on stops things being drawn outside the drawing area. The Zoom All command resizes the viewing area to the new limits.

Layers

Having decided on the size, it is now worth considering how your drawing is to be organised. Using AutoCAD's LAYER facility is the most efficient way of

Table 3.1 Some recommended layers in construction drawings

Field 1	Discipline	Field 2	Type of information
A	Architect	100–199	Ground, Substructure – General
B	Building Surveyors	200–299	Structure, Primary Elements
C	Civil Engineers	300–399	Structure, Secondary Elements
D	Drainage Engineers	400–499	Finishes
E	Electrical Engineers	500–599	Services – General
F	Facilities Managers	600–699	Services – Electrical
K	Client	700–799	Furniture and Fittings
S	Structural Engineer	900–999	External Works

Based on British Standard 1192 Pt5 and the Autodesk inc & AutoCAD
Users Group Layer Naming Convention for CAD in the Construction Industry

doing this. The layers of a drawing can be considered as a series of transparent sheets each containing parts of the drawing (Figure 3.2). Whole layers can be manipulated to change the color or linetype for all the objects on a particular layer. Layers can also be made invisible when they are not relevant to the current task, and then they can be made visible again later. For example, you might use one layer to contain all your construction lines, one layer for text and drawing margins, one for floor plan and another for the wiring diagram. Unlike a conventional paper drawing there is no need to delete your construction lines, simply make them invisible. In this way, when you come to edit the drawing or add in a plumbing diagram, all the original constructions are available.

Because of the importance of layers in the organisation of drawing information and the increasing need for drawing exchange via CAD, most users adopt some convention for using layers. A sample layer numbering standard is summarised in Table 3.1. For example, the layer S282 would be a layer created by the structural engineer for concrete columns. Readers are referred to the full standard for exact details. This highlights the importance of adopting a logical approach to organisation of information in CAD. Adherence to a common standard allows engineers using different CAD systems to communicate effectively.

When you start a new drawing in AutoCAD, it always has a layer called "0" (zero). Extra layers can be set up at any stage during the drawing but it is good practice to decide on the structure of the layers before commencing actual drawing. Releases 10 and upwards allow multiple colors and linetypes on a layer but still allow you to give a layer specific default settings.

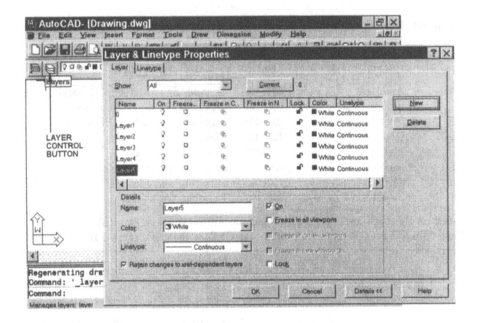

Figure 3.3 Layer Control Dialog Box

As an example, our new drawing will contain some doodling as you try out some drawing commands, some text, some construction lines and the tower outline. The following sequence gives the AutoCAD prompts and your responses.

Pick the Layer Control button shown in Figure 3.3 or pick **Format** from the menu bar followed by **Layer.**... This brings up the Layer Control dialog box, also shown in Figure 3.3. Initially, the layer name list on the left will contain only one value, namely 0. The button marked "Current" has 0 beside it indicating that layer 0 is the one to which new objects will belong.

To create the five new layers shown in Figure 3.4 pick the **New** button five times This generates layers with the names, "Layer1, Layer2" etc as shown in Figure 3.3. Our next task is to give them more descriptive names.

In the Layer Properties dialog box pick the text, "Layer1" so that it is highlighted. You can change the name by one of two methods. In the lower half of the dialog box there is an input field for the name. Double click in the field, delete the "Layer1" and type "CLINE". This will be the layer for the centerline of the tower. For the second method, pick the text, "Layer2". It becomes highlighted. Pick the highlighted **Layer2** once more. A small rectangle appears around the text and you can type in the new name, **CONST**. This method is

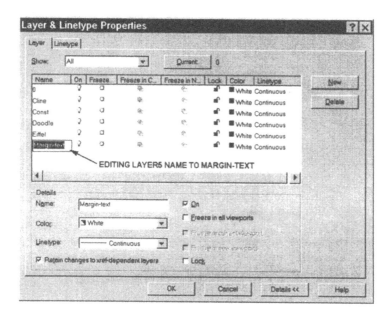

Figure 3.4 Layer Properties Dialog Box

similar to changing file names in Windows Explorer. When all the names have been changed to the ones given in Figure 3.4, click the **OK** button.

If your Layer & Linetype Properties dialog box is shorter than the one shown here, you may have to pick the button marked **Details >>** tp expand it. The column names may also be truncated e.g. you may see "F..." instead of "Freeze". Each column can can be resized by moving the mouse to the border of the column heading and draggin it left or right.

Layer names can contain letters and numbers and have up to 31 characters, though it is advisable to restrict them to as few as practicable. The names should be descriptive or follow some conventional numbering or acronym. AutoCAD will organize the list of layers in numerical and alphabetical order as soon as the OK button has been picked. Remember, even though it is good practice to plan our layers from the start, new layers can be added at any time.

When AutoCAD creates new layers they have a number of default properties. Their "State" is "On", that is, they are visible. They have a continuous linetype and the layer color for objects is white. Note that objects will only be white if your screen is configured to have a dark background. If you have a light background then "white" actually means black. White lines normally convert to black lines when plotted on paper.

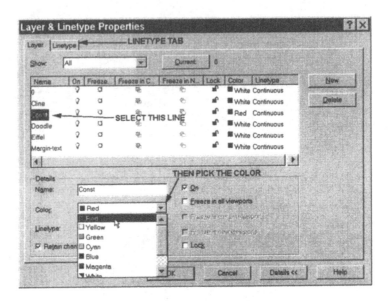

Figure 3.5 Select color for layers

As a first task let's give the Const layer the color red and the Doodle layer
the color blue. Move the arrow cursor into the Layer Name list and pick the
word "Const" on the line

Const ... White Continuous

The text "Const" should appear in reverse video (Figure 3.5). Now, pick the
color selection pull-down list in the dialog box. If the dialog box does not
display the pull-down list you will have to select the "Details >>" button.
You may have to scroll the list up a bit to find "Red". Click once to select
it. The color box and name should become updated in the dialog box. Red is
now the default or "Bylayer" color for the layer Const. Repeat this process to
make layer "Doodle" have Blue as its default color i.e. pick the line

Doodle ... White Continuous

Then pick the color pull down list from the dialog box and sleect **Blue**. Keep
the dialog box open as we haven't finished setting up the layers yet.

The next task is to select the linetypes for the Const and Cline layers. As
in the last chapter, we first have to load the appropriate linetypes. Let's try
"Dashed" and "Center" again. Pick the **Linetype** tab as shown in Figure 3.5.
Now pick the **Load** button and select the linetypes, **Center** and **Dashed** as we

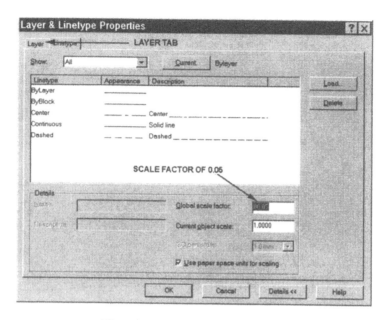

Figure 3.6 Scaling the linetypes

did in Figures 2.13 and 2.14 (pages 28 and 29). A quick way to scroll through the list of linetypes is to highlight any linetype, say, **Acad_iso02w100**. Then type **C**. This should zap you forward in the list to the first item beginning with "C", which in this case is "Center". While Center remains highlighted, scroll down a bit more until you see the "Dashed" linetype. Now hold down the **CTRL** key on the keyboard and pick **Dashed**. This allows you to select more than one linetype to load in one operation. Finally pick **OK**.

Your list of loaded linetypes should be the same as Figure 3.6. Move the cursor to the **Global scale factor** field and input the value **0.05**. Remember, the limits are now 65 by 45 while the base unit is one metre. This value of scale will make sure that the dashes are visible on the screen. If you want smaller or larger dashes just alter this scale factor. Keep the dialog box open for these linetypes to be assigned to the layers.

Pick the **Layer** tab at the top of the dialog box. Now pick the word "Continuous" on the line

Cline ... White Continuous

as shown in Figure 3.7. Now pick the **Linetype** pull-down list followed by **Center**. Similarly, give the Const layer the linetype "Dashed" i.e. pick the word "Continuous" on the line

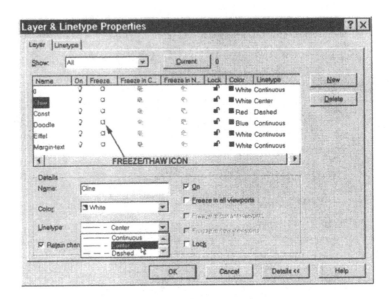

Figure 3.7 Setting layer linetypes

Const ... Red Continuous

and then pick **Dashed** from the list and pick **OK.**

Note: Each linetype that you load increases the size of yor drawing file by a small amount. If you load too many linetypes or create too many layers they can be unloaded by using the **PURGE** command. This command is described in Chapter 5.

The final layer operation is to set the layer MARGIN-TEXT as the current layer. This is done by selecting the "MARGIN-TEXT" on the line

Margin-text ... White Continuous

Then pick the **Current** button and then pick **OK** .

Note for Release 12 and 13 users: In earlier versions of the layer control dialog, you had to deselect layers before doing operations on other layers. In R14 layers are automatically deselected when another layer is picked. If you want to select more than one layer in the dialog, then hold down the CTRL key when picking the layers.

Layer control options

Of the other options in the Layer Control dialog box the most useful are the "Freeze" and "Thaw" icons (Figure 3.7). The icon appears as a sun when the layer is thawed and as a snowflake when it is frozen. Freezing a layer causes it to become invisible. AutoCAD ignores objects on frozen layers and so drafting can be speeded up by freezing layers when they are not required. Thawing a layer makes it visible again. Frozen layers are stored with the drawing when it is saved and can be thawed at any time. This feature will be used later in the chapter. Also useful is the "Lock/Unlock" feature introduced with Release 13. When a layer is locked, its objects are visible but cannot be edited. The lock icon appears as a padlock.

Layer property dialog box can be accessed using the Command line.

Command: **LAYER**

Note for R12 and R13 users: The layer command always brings up the dialog box when typed. If you want to use the R12 command line prompts use the old version of the command which is now called −**LAYER**. If you have scripts and Autolisp programs which make use of this, they may need to be altered.

Cursor location

If you move the cursor around the drawing area using the mouse the coordinates on the Status line change to give the current location. This is at the bottom left of the window (Figure 3.9). If the values do not change try pressing the pick button on the mouse. This will give the coordinate of that point. To switch on the continuous read-out of coordinates type **Coords** at the Command prompt.

Command: **COORDS**
New value for COORDS <0> **1**

There are three toggle methods also available: double clicking the coordinate display box; or ^**D** (press the CTRL key and while doing so press the D key); and the **F6** function key. Pressing any of these toggles the coordinate facility in a circular fashion between dynamic X,Y readout, static X,Y and dynamic polar coordinates. It is recommended that the dynamic X,Y coordinate display is switched on (ie COORDS value of 1). The dynamic polar display only operates when rubber banding is in operation eg drawing lines.

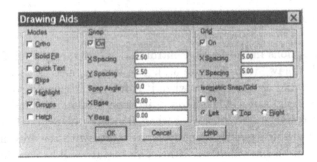

Figure 3.8 Drawing Aids dialog box

Rulers, grids and snapping

In addition to coordinate display AutoCAD gives a grid facility which helps you to get your general bearings within the drawing, and a snap function which helps to pin-point locations exactly. GRID causes an array of equally spaced dots to appear. These are analogous to placing a sheet of graph paper as a guide beneath the tracing sheet on a drawing board. The grid points are not part of the drawing and do not appear on plots.

The SNAP command is one of AutoCAD's most powerful (and time saving) facilities. By turning SNAP on you can restrict the movements of the cursor to discrete steps. The cursor jumps from one snap point to the next missing out all the points in between. This is particularly useful if you are drawing an item whose smallest dimension is, say 0.5 units. You could set SNAP to a value of 0.5. This makes point picking with the mouse a lot easier and eliminates the need to type the coordinates when exact locations are required.

The drawing, EXPRESS2, will be created with a SNAP value of 2.5. The GRID will be set to 5. This gives a nice relationship between the two settings. Every second snap point will also be a grid point. If the grid is too dense it can get in the way and slow down the graphics. The best way of setting this up is to pick **Tools** from the menu bar and **Drawing Aids...** from the pull-down menu. This gives the Drawing Aids dialog box, Figure 3.8.

The middle section of the dialog box concerns the SNAP settings. Move the cursor to the box beside **X Spacing** click and overtype the current value with **2.5**. If you click on the **Y Spacing** box it will automatically update to be the same as the X value. Make sure that the **On** box is ticked. If it is empty pick the On box with the cursor to activate SNAP. When inputting the snap size, make sure to include the feet and inches symbols. Similarly set **Grid** to

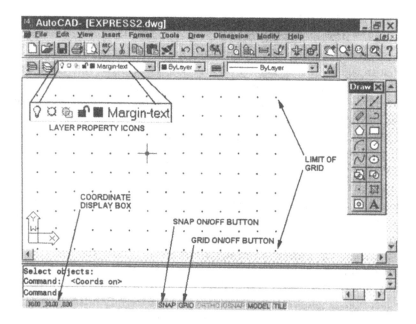

Figure 3.9 Snap and grid

be **On** with X and Y Spacing of **5**. When your box matches Figure 3.8 pick **OK**. The display should then look like Figure 3.9.

When using these features, try to choose sensible spacings. If the spacing is small, the redraw speed will slow down, as AutoCAD has to re-display all the dots and ticks. If the spacing is too small you will get a message "Grid too dense to display". If this happens just re-issue the command and set new spacings.

The Windows status bar in Figure 3.9 shows the SNAP and GRID on/off toggle buttons. **Double clicking** either of these switches the mode. When the function is inactive the text is greyed out.

The function key **F9** also toggles SNAP while **F7** works for GRID. Release 12 users can still switch SNAP by pressing ^B and ^G works for GRID.

The GRID has a useful side effect in that it shows the limits of the drawing when Zoom All has been activated. This is particularly handy in Windows where you can have oddly proportioned drawing areas.

It doesn't matter which layer you are on when activating GRID or SNAP. They act as an overlay on the whole drawing. They can be switched on and off as many times as required and can be used independently of each other. Remember, the GRID will not appear on your print out or plot.

Two Dimensional Coordinate notation

You have already used the X,Y coordinate notation. In the last chapter the X
was seen to signify the horizontal distance from the origin and Y the vertical
distance. A point on the drawing can be located by keying in a pair of numbers
separated by a comma. For example, the coordinates "4,3" belong to a point 4
units to the right and 3 units above the origin. The origin has the coordinates
"0,0". Mathematics books call this the Cartesian coordinate system after the
French philosopher and mathematician, Descartes. AutoCAD calls it the X-
Y WORLD system (WORLD meaning that the values are in relation to the
drawing origin).

Sometimes it is more convenient to work in distances relative to one's
current location. For example, when giving directions to a stranger in town
your instructions might be "Follow this road for two miles and turn right. The
Computer Training Firm's offices are a further 100 m on the left." This is a
lot more meaningful than giving the location in terms of a map grid reference.
Similarly, you can tell AutoCAD to draw a line giving directions relative to the
cursor location. The notation used in AutoCAD to signify *relative* coordinates
is to put "@" in front of the X,Y pair of values. For example to draw margins
for EXPRESS2 try the following:

Command: **LINE** or pick the Line button from the floating toolbar.
From point: **2.5,2.5**   (A)
To point: **@60,0**   (B)
To point: **@0,40**   (C)
To point: **@−60,0**   (D)
To point: **close**

The point A must be in absolute X-Y WORLD coordinates since it is the first
point to be input to the drawing. The second point, B, is given as 60 m to
the right of A and at the same height. C is 40 m directly above B and so on.
Relative coordinates always use the immediately preceding cursor location as
a temporary origin.

If you know the length of a line and the angle it makes with the horizontal
then it can be easier to use *relative polar* coordinates. To draw a vertical centre-
line from point E (30,5) which is 36.5 long first switch layers and use the LINE
command again.

The easiest way to switch layers is to pick the **down pointing arrow** on
the right of the current layer box on the tool bar, Figure 3.10. This gives a
pull-down menu listing all the available layers. The icons in front of the names
signify some of that layer's properties. All the layers are currently on, unfrozen
(thawed) and unlocked i.e. they are all visible and accessible. The color of the
small square indicates the default color for that layer. Pick **CLINE** and the

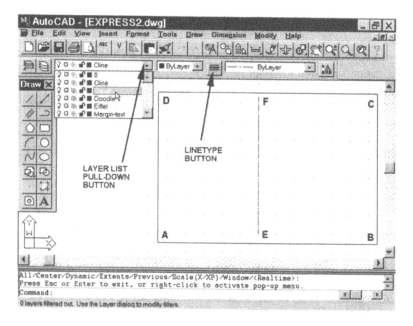

Figure 3.10 Margins and center-line

change is executed. The Current Layer and should change to CLINE and the line type also changes.

Having switched to the CLINE layer, now draw the line from E to F.

Command: **LINE**
From point: **30,5**   (E)
To point: **@36.5<90**   (F)
To point: **<ENTER>**

The "<" or "less than" character indicates that the next character is an angle. Thus the above line will be drawn at 90 degrees from the horizontal. As positive angles are anti-clockwise the line goes up and not down. Again the "@" means that the location is relative to the previous point (30,5). If you leave out the "@" you will be using *absolute polar* coordinates. Then the angles relate to lines from the origin to the new location. This can be cumbersome and is not generally recommended.

Did the center line come out dashed? If the line appears continuous, this is a symptom that the linetype scale is not quite right. Remember, the value of "LTSCALE" controls the lengths of the dashes and gaps. The actual value you use will depend on the limits of your drawing and the paper size you use for plotting. If Linetype Scale is too small then the gaps between the dashes will be imperceptible. Here you need to set a value of 0.05 as outlined earlier on page 45.

If you wish to change the linetype scale setting, pick the Linetype button from the properties toolbar (Figure 3.10). With the mouse, highlight the Global Linetype Scale and change it to 0.05 or whatever value works for your screen (Figure 3.6, page 45). This scale factor applies to all linetypes on all layers. To see the effect you might have to regenerate the drawing.

> Command: **REGEN**
> Regenerating drawing.

If you want to experiment with relative and absolute coordinate notation, try drawing some objects on the DOODLE layer. On this doodle detour of the AutoCAD Express you can take in some sights on the scenic route. If you wish to press ahead without further practice you can skip forward to the next section entitled "Digital setsquares".

The pyramids of Egypt can be drawn as triangles (Figure 3.11). First, switch to the DOODLE layer. This can be done using the pull down list as before bar or by typing −LAYER at the Command prompt.

> Command: **−LAYER**   (Note the minus sign before Layer)
> ?/Make/Set/...: **S**
> New current layer <0>: **DOODLE**
> ?/Make/Set/...: **<ENTER>**
> Command: **LINE**
> From point: **10,30**   (P1)
> To point: **@5,6**   (P2)
> To point: **@5,−6**   (P3)
> To point: **@10<180**   (P1)
> To point: **<ENTER>**
> Command: **LINE**
> From point: **5,15**   (P4)
> To point: **@12<60**   (P5)
> To point: **@12<−60**   (P6)
> To point: **close**

If you have the screen menu at the right hand side of the drawing area, the **close** option can be picked. Otherwise you will have to type **C**. Experiment with other shapes and combinations of the different coordinate notation.

Now is as good a time as any for saving the drawing. Remember the motto "Save early and often". Pick the Save button from the Standard toolbar (the icon that looks like a floppy diskette) or pick **File/Save**. This executes the quick save command.

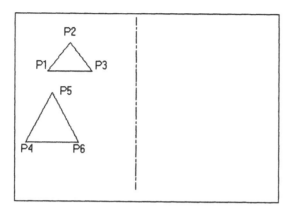

Figure 3.11 The pyramids of Egypt

Command: **QSAVE**

Digital setsquares

A large part of any drawing consists of horizontal and vertical lines such as
the margins and centre-lines above. AutoCAD's ORTHO command allows you
to draw these horizontal and vertical lines quickly and with 100% accuracy.
When the ORTHO mode is ON then all movements of the cursor are restricted
to either the X direction or the Y direction.

The simplest way to switch ORTHO on is to *double click* the **ORTHO**
button on the Windows status bar (Figure 3.13). Pressing the **F8** function key on the keyboard also toggles ORTHO. It can also be accessed by
Tools/Drawing Aids... or at the command line.

Command: **ORTHO** ON/OFF: **ON**

Using one of the toggles is easier and pressing it a second time switches back
to normal drawing mode. When it is active the word "ORTHO" appears black
on the status line at the bottom of the screen. The command prompt area will
echo your action with "<ORTHO ON>". To examine these effects draw the
horizontal construction lines outlined below for EXPRESS2 making sure that
ORTHO is ON.

Use the command line or Layer pull-down list to set the **Current** layer
to CONST and to **Freeze** the DOODLE layer. To freeze the layer pick the
"sun" icon on by DOODLE and change it to a snowflake as in Figure 3.12.
Then pick the **CONST** layer from the list.

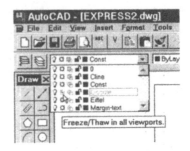

Figure 3.12 Freezing the Doodles

You are now ready to draw some construction lines. Lines of "infinite" length can be drawn using the XLINE command. This is really neat for making construction lines. The **XLINE** command can be executed by picking the **Construction Line** button on the Draw toolbar (the double headed arrow button in Figure 3.13. You can also pick **Draw/Construction Line** from the Menu bar. Initially we will define the line by picking two points.

> Command: **XLINE**
> Hor/Ver/Ang/Bisect/Offset/<From point>: **17.5,5**   (G)
> Through point: pick a point to get a horizontal line.
> Through point: <**ENTER**>

Now move the cursor to the right of the centre-line (EF) near to G'. Even though the cursor is not exactly horizontally across from the point G, the line from it is (Figure 3.13). If you move the cursor back towards G there is a point when the line suddenly jumps to being vertical. So, wherever you pick the point the line is restricted to the X and Y directions. The governing factor for whether it is vertical or horizontal is the larger magnitude, X or Y, from the first point to the cursor. Move the cursor back to G' and pick the through point so a to to have a horizontal line.

You may have noticed that one of the options with XLINE is to draw horizontal construction lines. The rest of the lines on Figure 3.14 will be drawn with this method.

> Command: <**ENTER**> or pick the Construction Line button.
> **XLINE**
> Hor/Ver/Ang/Bisect/Offset/<From point>: **HOR**
> Through point: **0,15**   (HH')
> Through point: **0,17.5**   (JJ')
> Through point: **0,22.5**   (KK')
> Through point: **0,25**   (LL')

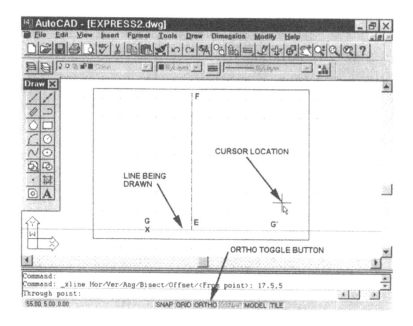

Figure 3.13 ORTHO mode construction line

Through point: **0,38**   (MM')
Through point: **<ENTER>**

Your drawing should now resemble Figure 3.14 (without the letters). Your construction lines should be red.

Now is another good time to save your drawing. You can save it as many times as you like, during a session.

Command: **QSAVE**

Snapping to objects

The main usefulness of construction lines is that by drawing them you reduce the number of calculations necessary to locate awkward points. The intersection point between a line and a circle is easy to draw but may require clever geometry to calculate. This is not to suggest that the AutoCAD user is not clever. Rather, you just want the fastest and simplest solution!

AutoCAD allows you to snap to key points of previously drawn items. Pick **View/Toolbars** followed by **Object Snap** to bring up the Object Snap floating toolbar (see Figure 3.15). You may need to scroll down to find the

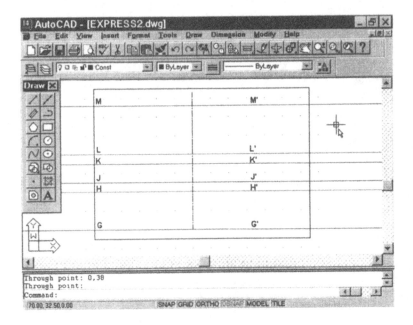

Figure 3.14 Horizontal construction lines

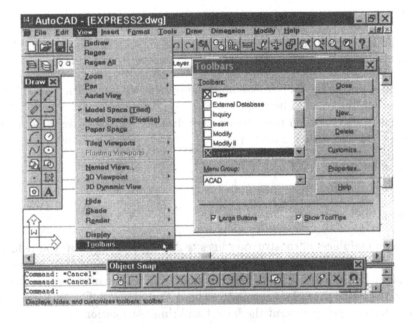

Figure 3.15 Object Snap menu and buttons

Object Snap box to check. Pick **Close** to remove the Toolbars dialog. Finally, drag the Object snap toolbar to a convenient location. In Figure 3.16 the toolbar is locked on the right.

The Object Snap buttons give you facilities to locate the centre or tangent point of a circle or arc, the end point and mid point of a line, the insertion point of text etc. These snapping features will be used to draw the outline of the Eiffel Tower

First, draw a line from G that intersects JJ' at an angle of 58 degrees. Relative coordinates are used below. Then use object snap to locate the intersection point N for the start of the subsequent line.

Command: **XLINE**
Hor/Ver/Ang/Bisect/Offset/<From point>: **Ang**
Reference/<Enter angle (0.0)>: **58**
Through point: **17.5,5**   (G)
Through point: **<ENTER>**

We will now use Object Snap in single point mode. We can use the Object Snap buttons whenever AutoCAD requests a point (eg "From point:"). The next few lines illustrate this feature. Type **LINE** and at the "From point:" prompt pick the **Intersection** button in the toolbar (the one that looks like an X in Figure 3.15). The prompt changes to "_int of:". Move the cursor until you are near point N on the line JJ'. Once it is close enough, a small X icon will appear at the intersection point. This tells you that AutoCAD has found an intersection point. Now press the pick button on the mouse to select the point N. The line from N is to intersect KK' at 77 degrees. The length of the line is not important, as long as it intersects KK'.

Command: **<ENTER>**
LINE From point: int of: pick point N
To point: **@5.5<77**
To point: **<SPACE BAR>**

Remember, the space bar acts just like the enter key. Now to draw the vertical line on Figure 3.16 from the point, P, and perpendicular to the line MM', use toolbar "Snap to Perpendicular" button on bottom line. Remember, the Drawing Aids dialog box was used to set a snap grid of 2.5. This may make it difficult to pick the line MM'. To get around this turn off the snap by pressing F9 function key or double picking "SNAP" from the status bar. This has no effect on Object Snap.

Command: **<SPACE BAR>**
LINE From point: **28,25**   (P)
To point: **PER** to F9 <Snap off> Pick line MM'

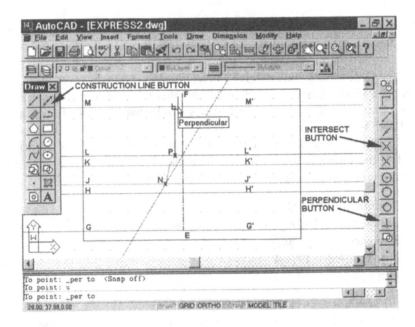

Figure 3.16 Using Object Snap options

To point: ^**B** <Snap on> <**ENTER**>

This way of using Object Snap, gives a single point selection and then returns to normal selection straight away. It can be used only when AutoCAD is expecting you to pick a point.

When you want to connect up a lot of construction points using Object Snap it can become tedious to have to pick, say, Intersection each time. The second way of using Object Snap is to set up a continuous "Object Snap" mode. This is done by picking **Tools/Object Snap Settings**... from the menu bar. Alternatively, type DDOSNAP at the command line. This brings up the Object Snap settings dialog box, Figure 3.17. Pick **Intersection** and then **OK**.

This means that AutoCAD will always snap to the intersection point nearest to the centre of the target box. Note that Figure 3.17 also has a section for controlling the "Aperture Size" or the size of the target box. The slider bar can be used to either increase or decrease the size. The smaller the aperture the faster AutoCAD will locate the point. However, the smaller size makes it more difficult for the user to pick objects. Picking the AutoSnap tab gives the Autosnap settings allowing you to alter the color of the marker, specify

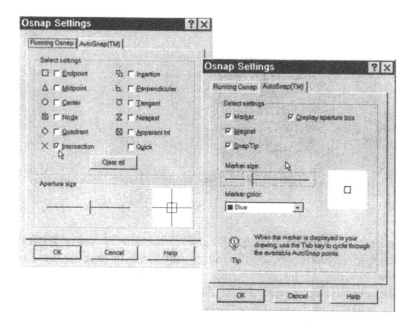

Figure 3.17 Running Object Snap settings

whether the aperture box is displayed or not. Note that the AutoSnap display aperture box must be checked if you want to see the aperture.

Now, use this to draw the outline of the tower (Figure 3.18). Note that, for clarity, the construction lines in the figures have been altered to appear dashed. First, change layers by picking the **Layer list pull-down** next to the layer name box on the tool bar and then pick **EIFFEL**.

Command: −**LAYER**
?/Make/Set/...: **S**
New current layer <0>: **EIFFEL**
?/Make/Set/...: <**ENTER**>
Command: **LINE:**
From point: **22.5,5**   (Q)
To point: pick the point, G where GG' intersects GN   (G)
To point: Pick R (Note: Object snap overrides ORTHO)
To point: <**ENTER**>
Command: <**ENTER**>
LINE From point: Pick S
To point: Pick R

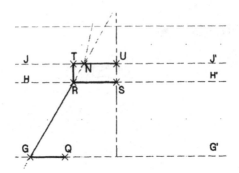

Figure 3.18 OSNAP mode INTersec

Now to draw a line from R perpendicular to the line JJ' select **PERPend** from the Object Snap menu. This causes AutoCAD to look for the nearest point that is either an intersection point or a normal (perpendicular) to JJ'. To make sure that the perpendicular line is drawn pick the line JJ' far away from any intersection points. Once the point T has been picked the object snap reverts to only looking for intersections.

> To point: perp to: Pick line JJ'   (T)
> To point: Pick U on Figure 3.18
> To point: <**ENTER**>

This procedure is repeated to draw the remainder of the left hand outline (Figure 3.19).

> Command: **LINE**
> From point: Pick N
> To point: Pick V
> To point: <**ENTER**>
> Command: <**ENTER**>
> LINE From point: Pick W
> To point: Pick V
> To point: **perp** to: Pick line LL'   (X)
> To point: Pick Y
> To point: <**ENTER**>
> Command: <**ENTER**>
> LINE From point: **28,25**   (P)

If you have difficulty in selecting P eg AutoCAD snaps to the nearby point, X, then the target box aperture is probably too large. Cancel the LINE command and use DDOSNAP (Figure 3.17) to reduce its size and repeat the line PZ operation.

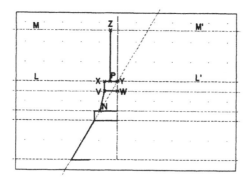

Figure 3.19 Outline of the Eiffel Tower

To point: Pick intersection point on MM'   (Z)
To point: <**ENTER**>

Now to turn the OSNAP mode off. Pick **Tools/Object Snap Settings**
and then click the **Intersection** box to remove the X or pick **Clear All** and
pick **OK**. An alternative way of doing this is to use the OSNAP command as
follows.

Command: **OSNAP**
Object snap modes: **NONE**

As you pick points on the screen, little blips or crosses may appear. The blip
gives you some visual feedback about the location of the point and it is easy
to see if it is at the correct location. However, the more points you pick the
greater the distraction caused by these blips. To clear the screen of old blips you
must execute the **REDRAW** or **REGEN** command. This simply redraws the
current display and as the blips are not true parts of the drawing they do not
reappear. REDRAW is in the Dos VIEW pull-down menu and also appears
as the Pencil button in the standard Windows toolbar. If you really hate
these blips appearing, you can turn them off for good. Pick **Tools/Drawing
Aids...** from menu bar and remove the check from the "Blipmode" box in
the dialog box.

Command: **REDRAW**

ZOOM and PAN

You have already used the "ZOOM All" command to display the whole drawing
area. In this section you will use ZOOM to enlarge the view of the top of the

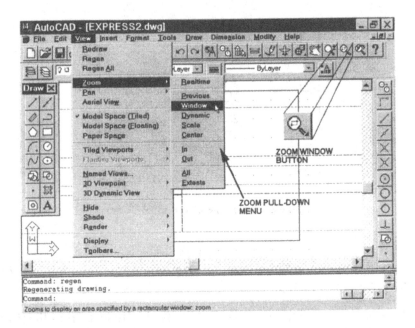

Figure 3.20 Zoom pull-down menu and button

tower being drawn so that greater detail can be added to it. The best way
to imagine how zooming works is to think of the computer display as the
image you would see looking through a camera lens. As you adjust your zoom
lens the item you are looking at becomes enlarged while peripheral items are
excluded from view. The PAN command is another term from the camera
man's vocabulary. It allows you to sweep your "camera lens" over the drawing
to look at other parts with the same magnification.

To enlarge the top of the tower pick **View** from the menu bar and then
ZOOM. Then pick **Window** from the ZOOM menu, Figure 3.20. You are
asked to give the two opposite corners of the new window of vision. There is
also a ZOOM-Window button on the standard tool bar.

Pick the points W1 and W2 shown on Figure 3.21.

Command: '_zoom	
All/Centre/.../<Scale(X/XP)>: _window	
First corner: **25,35** or nearby	(W1)
Other corner: **38,44**	(W2)

The display should now change to give a close-up view of that rectangular
window (Figure 3.22). The spacing between the grid points and axis ticks
looks larger but the coordinates are just the same. As you move the cursor

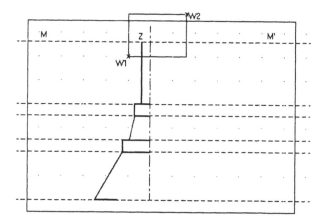

Figure 3.21 Picking a window

around the drawing the read-out on the status line will give the coordinates with respect to the drawing origin. Nothing has changed in the drawing, it's just the amount you can see that has altered.

If your window is significantly different from that in Figure 3.22 then the wrong window may have been chosen. Do a ZOOM All and repeat the ZOOM Window, typing in the coordinates for W1 and W2.

As you are working at a larger magnification to draw small items you can change the GRID and SNAP settings. This can be done with **Tools/Drawing Aids. . .** or by typing the commands. If you type the GRID command there is an extra facility available.

Command: **SNAP**
Snap spacing or ON/OFF/Aspect/Rotate/Style <2.50>: **0.5**
Command: **GRID**
Grid spacing or ON/OFF/Aspect/Snap/Style <5.00>: **2X**

The "2X" sets the grid to twice the current snap setting. You don't get this option with the dialog box.

Now draw the viewing platform with its roof and antenna (Figure 3.22). Use the Object Snap **Intersection** button on the tool box for points A and Z.

Command: **LINE**
From point: **INT** of: Pick A
To point: **@−2.5,0**   (B)
To point: **@0,−1**   (C)
To point: **@2.5,0**   (D)

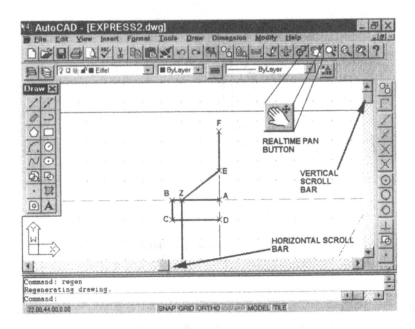

Figure 3.22 Half the viewing platform

To point: <**ENTER**>
Command: <**ENTER**>
LINE From point: **INT** of: Pick Z

In the next operation, you can switch ORTHO off in the middle of the LINE command by pressing **F8** or double clicking the ORTHO and then pick the point E (30,39.5), making use of the new snap settings. Alternatively, you can type the coordinates (typing coordinates overrides ORTHO).

To point: <ORTHO OFF> **30,39.5**   (E)
To point: **@0,2**   (F)
To point: <**ENTER**>

Now ZOOM in even closer using window to add some structural detail. Pick the **ZOOM-Window button** from the tool bar. The alias quick key for ZOOM is **Z** for keen typists. Then type the coordinates given below.

Command: **Z**
ZOOM
All/Centre/.../<Scale(X/XP)>:**W**
First corner: **27,35.5**

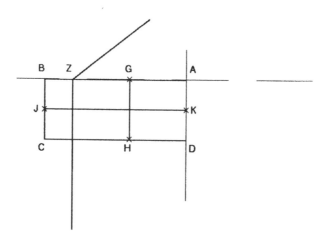

Figure 3.23 Adding some structure

Other corner: **30.5,39**

This increases the magnification. Now, using the same SNAP resolution, you can draw the vertical line GH and the horizontal line JK, as shown in Figure 3.23.

```
Command: LINE
From point: 29,38                                              (G)
To point: @0,−1                                                (H)
To point: <ENTER>
Command: <ENTER>
LINE From point: 27.5,37.5                                     (J)
To point: @2.5<0                                               (K)
To point: <ENTER>
```

To ZOOM back to the last magnification pick the Zoom Previous button (beside the Zoom Window button) or pick **View/Zoom** and then pick **previous** from the menu. Alternatively, type:

```
Command: Z
ZOOM
All/Centre/.../<Scale(X/XP)>: P
```

AutoCAD stores the settings of up to ten previous zoomed views and so you can step backwards that many times. This facility saves you having to do a **ZOOM All** followed by a new ZOOM Window to get back to an earlier view. Your picture should now resemble Figure 3.22 but with the new lines in place.

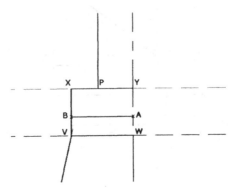

Figure 3.24 Middle landing after PAN

When you are happy with the magnification but need to access part of the drawing which is now off screen, use the PAN button or the horizontal and vertical scroll bars indicated in Figures 3.22. PAN is also available on the View pull-down menu. For example, to see the middle landing of the Eiffel tower under construction try to use the vertical scroll bar. Pick the "down arrow" at the bottom of the scroll bar to move the view. You can also use the scroll button to move more quickly but this can be tricky at high magnifications. If you use the PAN button, the cursor changes to a hand shape. Press the left mouse button to make the hand "grab" the image and drag it around. Release the mouse button to "let go" of the image. Then press the **Esc** key to exit PAN.

Hopefully, you can adjust your screen image to show the middle landing part of the tower structure as shown in Figure 3.24. If this proves a little difficult, do not dispair. Panning can take a little practice. There will be more about zooming and panning in the next chapter.

As a last resort, you can get to the desired display by executing Zoom window with the coordinates of the window (23,20) to (36,29). This middle landing of the tower needs one more line. Draw the line AB using Object Snap perpendicular (Figure 3.24).

A note for R12 users: Typing PAN always executes the realtime version. If you need to use the R12 version of the command you will now need to type −PAN (minus pan).

Command: **LINE**
From point: **30,23.5**   (A)
To point: **PERP** perpend to pick line VX   (B)

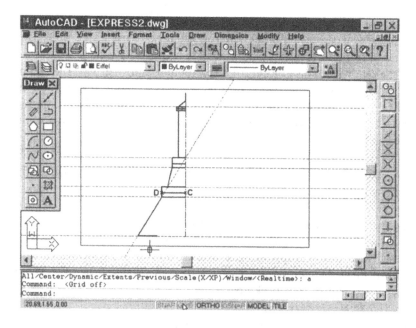

Figure 3.25 Back to full view

To point: **<ENTER>**

To see the lower landing pick the **PAN** button once more and drag the image up the screen. Again, if you have difficulty in locating the landing either zoom out fully or zoom Window to (20,12) and (35,20)

> Command: '_pan
> Press Esc or Enter to exit, or right click to activate pop-up menu
> **Esc** after dragging the image.
> Command: **LINE**
> From point: **30,16**   (C)
> To point: **PERP** perpend to pick vertical line at D (Figure 3.25)   (D)
> To point: **<ENTER>**

Finally, do a ZOOM-All.

> Command: **Z**
> ZOOM
> All/Centre/.../<Scale(X/XP)>: **A**

and the display should match Figure 3.25.

The picture may not look like much at the moment, but if you skip ahead
a few pages you can see its potential. You will need to use EXPRESS2 again in
Chapter 5, so make a safe backup copy. Do a Quick Save by picking the Save
icon (the one with a disk on it) from the toolbar and then *type* the command
Save. This brings up the "Save As" dialog box. Overwrite the file name with,
say, **TOWER**. Make a copy onto a floppy disk as well. Place a formatted
diskette in Drive A: and do another **SAVE** selecting the A drive from the
pull-down "Save in" list at the top of the dialog box.

HAZARD WARNING! In the unlikely event that you are using the floppy
disk drive to store your original drawing files do not remove the floppy disk
at any time during the editing session. The default AutoCAD configuration
stores temporary files in the same directory as the original drawing. If you
remove the disk before exiting AutoCAD then these temporary files will not
be erased and may corrupt the disk. When you exit AutoCAD from File the
pull-down menu, all temporary files are deleted. You can use the Windows
Explorer to organise your back ups.

Note about current name of drawing: If you use the Save As option in the File
pull-down menu, AutoCAD switches the name of the working drawing to the
new filename, i.e. the name of the drawing on the window title bar changes.
However, if you type the SAVE command and change the name, the working
drawing file remains the one you originally saved or opened.

Keyboard toggles and transparent commands

Before ending AutoCAD for this chapter, here are some descriptions of the
toggles that can speed up your drafting. By pressing the CTRL key in combination with other keys, various commands can be executed, even in the middle
of doing another command. Some of the toggles cause alterations to the current command. Some of the keyboard function keys duplicate these. Table 3.2
gives a list of the main function keys and keystroke combinations.

A transparent command is one that can be run while another is still being
executed. For example, the ORTHO, F8, command was issued above in the
middle of a LINE sequence. This type of transparency of the ORTHO command
greatly increases the flexibility with which it can be used. Similarly SNAP and
GRID can be used transparently by pressing the toggle, CTRL code, function
key or Drawing Aids menu item. The single point usage of Object Snap is
another example.

REDRAW, ZOOM and PAN can run during other commands if they are
picked from the View pull-down menu or if they are prefixed by a ' (eg Command: 'ZOOM etc.). The 'ZOOM and 'PAN will only work if the change in

Table 3.2 Keyboard Toggles

Key	Action
F1 or ?	Executes AutoCAD's Help facility
F2	Toggles text and graphics windows
F3	Object Snap settings
F4	for use with tablets
F5	for use with Isometric drawing - see Chapter 8
F6 or ^D	Dynamic - static cursor location read-out
F7 or ^G or Status bar	Toggles the GRID on and off
F8 or Status bar	Toggles ORTHO on and off
F9 or ^B	Toggles SNAP on and off
^C	Copyclip command - (this is a change from R12)
^O	Opens a new file (this is a change from R12)
Esc	Causes pull-down menus to disappear
Esc	also Cancels the current command

magnification is less than a factor of about 10. If it is more than 10 then AutoCAD will probably have to REGENerate the drawing to maintain sufficient screen resolution.

Another transparent command that is useful to know about is the context sensitive help facility. If you are in the middle of a command such as LINE and you want to find out what to do next type **'Help** at any time.

Command: **LINE**
From point: **'HELP** or **'?** or **F1 key**

To exit AutoCAD Help press **Esc**. When you exit the help facility you can continue with the command or cancel it. A second Esc is needed here if you want to cancel the line command.

Resuming LINE command
From point: **Esc**

This HELP information is intended as a reference only and not as a learning aid. The explanations of commands are thorough and so can appear complicated. Use AutoCAD Express for learning new commands and use HELP as a reminder. In this way you will understand the workings of the commands and be relieved of the overhead of having to remember everything. The online help, 'HELP, is also available through the pull-down menu and appears as a button on dialog boxes.

One more command that can be executed transparently is LAYER. However, not all of the options within LAYER can be used in this way. Pick the Layer control button, the Layer pull-down list or **Format/Layer...** pull-down menu from the menu bar. Some options such as thawing a layer or changing its linetypes will only be executed when the drawing is next REGENerated and so will not come into force immediately.

Finish for the time being

The final task in this chapter is to exit AutoCAD for a well deserved break. Pick **File** and **Exit**. Pick **Yes** in the Save Changes dialog box if it appears. One of the chief purposes of exiting any application is to remove all the temporary files from the hard drive and to return the memory to the system. Long CAD sessions can lead to a build-up of temporary files which are only purged when you exit AutoCAD.

This is not the end of the EXPRESS2 drawing. The "Exit" just means to end this editing session. You will return to edit it and add in some fancy iron work in Chapter 5. So don't erase it from your disk! And keep a safe backup copy on another disk.

Summary

In this chapter you have become an expert LINE drawer and have encountered most of AutoCAD's drawing aids. When doing an AutoCAD drawing, you need to think in terms of these facilities so that your plan of action makes best use of them and speeds up your drafting. Always start drawings with the key construction lines.

You should now be able to:

Set up layers and alter their colors and linetypes and Freeze a layer.
Input relative, absolute and polar coordinates for points.
Set up a suitable grid overlay.
Restrict cursor movements to discrete SNAP points.
Restrict cursos movements to ORTHOgonal directions.
Locate the intersection of lines.
Draw lines perpendicular to other lines.
Draw infinitely long construction lines.
ZOOM in to magnify details and ZOOM out to see the whole picture.
PAN across the magnified picture.
Make back up copies of your drawing.
Run commands transparently.

Chapter 4 DRAWING AND EDITING

General

The AutoCAD Express takes to the air for this chapter's exercise. The bulk
of AutoCAD's drawing objects are introduced along with some simple editing.
This express air-service will be by hot air balloon (Figure 4.1)!

Figure 4.1 Target drawing for chapter 4

The first part of this exercise will look at erasing objects and setting up the
correct layers. We will be using much the same drawing environment as EX-
PRESS1.DWG from Chapter 2. We will open this drawing file and save it
under a new name. Then we will delete some objects before drawing our hot
air balloon. Make sure that you have a copy of your previous drawing EX-
PRESS1.DWG.

If you don't have a copy of EXPRESS1.DWG handy you could make one
up. It should have limits of (0,0) and (65,45) and set an LTSCALE value of
0.1. Now draw some lines. It doesn't matter where or how many but you will
have to improvise when erasing them later. Add your name using DTEXT at
the point (40,10) with a height of 2 units, zero rotation and then SAVE the
drawing with the name **BALLOON**.

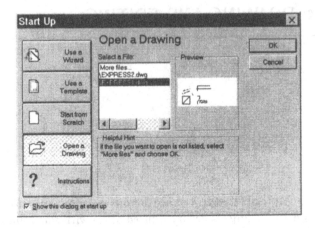

Figure 4.2 Opening an existing drawing

Opening a drawing file

There are two ways in which you might open an existing drawing file. Firstly, when you start AutoCAD, the Start up dialog appears. Pick the button marked **Open a Drawing** as shown on Figure 4.2. AutoCAD stores the names of the most recent files. If you highlight one of these files with one mouse pick, a thumbnail sized image of the file appears in the preview box. Double clicking on the file name or picking **OK** opens the highlighted file.

If EXPRESS1.dwg does not appear in the list of recent files, double click on **More Files** at the top of the list. This brings up the Select File dialog box as shown in Figure 4.3. You can navigate around your disk using this dialog box in the usual Windows manner. Once you have found your file, one mouse pick shows the preview. Picking the **Open** button then brings the file into the drawing editor.

When AutoCAD is already running and you wish to open a file, you can use the normal Windows method i.e. pick **File/Open** from the menu bar. This also brings up the Select File dialog shown in Figure 4.3. Note that the File menu also contains the most recently accessed filenames.

To change the drawing name pick **File/Save As...**, type **BALLOON** in the filename field and pick the **Save** button. The name on the title bar at the top of the screen should now have changed to BALLOON.dwg. The screen should now appear like Figure 4.4 which should be the same as Figure 2.15 on page 30. Any edits we make will affect only the BALLOON.dwg file. The file, EXPRESS1.dwg is now safely closed and stored on the disk.

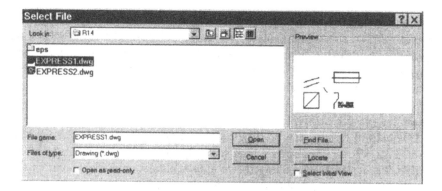

Figure 4.3 Select file dialog box

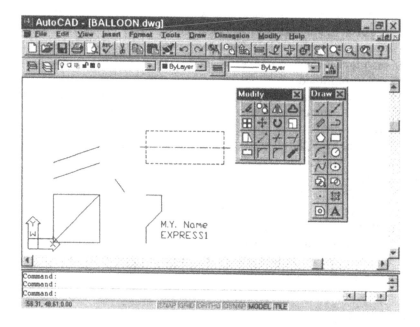

Figure 4.4 BALLOON.dwg initial state

Rubbing out and OOPS

When something has been drawn incorrectly or is no longer required on the drawing you can rub it out. This task is a lot easier with AutoCAD than on a conventional drawing board. With AutoCAD's ERASE command you will always get perfect results. The first part of the Balloon exercise is to selectively delete all the objects in the drawing. This will be done in stages to introduce the "Object Selection" procedure. Once this is accomplished, you can begin to draw the picture in Figure 4.1.

It is helpful to have the Modify toolbar visible as it contains the ERASE button. Pick **View/Toolbars...** and scroll down the list of toolbars to select **Modify**. Then pick **Close**. Drag the Modify toolbar to a suitable part of the screen so that it does not obstruct the drawing objects.

Object selection modes

Objects, lines, text, etc. can be selected for editing in one of two modes. Users of earlier versions of AutoCAD will be familiar with the "verb/noun" mode. The command or "verb" is picked first and then the objects, or nouns to be edited are selected. For example, to erase a line one would first pick the ERASE command and then pick the line. Since the launch of AutoCAD Release 12 you can use the more efficient "noun/verb" mode. You pick all the objects first and then pick the command.

To make sure that noun/verb selection has been enabled pick **Tools** and **Selection....** In the dialog box, Figure 4.5, make sure that Noun/Verb and Implied Windowing have an "x" in their boxes (ie are enabled) and click on the **OK** button. The default setting should have these enabled. Note that this dialog box also allows you to change the size of the pick box that is used for selecting objects.

The way lines or other objects are selected for erasure is flexible. You can place the cursor on the item to be scratched and press the pick button, move to the next item and pick it and so on. You can select a whole group of objects using a window like in the ZOOM command or you can erase just the last object that was drawn. Indeed, the object selection procedure can even use a combination of all these methods.

Erasing some lines

Firstly, you will erase two of the lines by simply picking them (Figure 4.6) and then picking the Erase button on the tool box. Move the cursor to the lines and pick the points P1 and P2. When the line is successfully selected it becomes ghosted. 3 little boxes, called grips, may also appear when the line is selected.

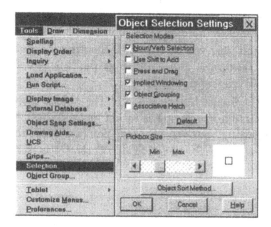

Figure 4.5 Enabling Noun/Verb Selection mode

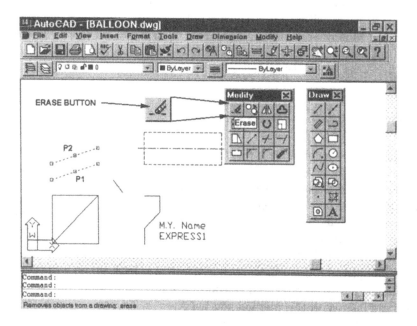

Figure 4.6 Erasing two lines

The ghosting and grips give a visual acknowledgement of the selection. The grips can also be used in some advanced editing procedures. Now pick the **Eraser button** from the tool box. The command line will echo that 2 objects have been found and they will be erased. If the grips did not show up on your screen pick **Tools/Grips...** and **Enable grips.** It is a good idea to have the grips enabled as we will see in Chapter 5.

If you press the mouse button but miss the object, it won't go dotty and the cursor will go into a windowing selection method. Note that on Figure 4.5 the "implied windowing" box was checked. You can cancel the window by **Esc** and resume picking. The next deletion demonstrates how to use windows to select objects.

Window selection

With window selection the order and relative position of the two corners is significant. For example, the window picked from W1 (**30,2**) to W2 (**47,25**) shown in Figure 4.7 finds the four lines, ghosts them and displays their grip points. Now pick the **ERASE** button and they disappear. Note that the text which is partially within the window is not selected. Only items that are fully within the window are selected. If any part of an object is outside the window it will not be erased. For this mode of window selection, the first point picked must be to the left of the second as in this example.

OOPS! I didn't mean to rub that out

If unexpected items disappear from the display, first try a **REDRAW**. If one item overlaps another and one of them is erased it looks like all the overlapped section has been deleted. Executing a **REDRAW** corrects this display error. If the missing item doesn't reappear then it probably has been deleted. Type U at the command line to undo the last command. Alternatively, pick the UNDO button from the toolbar at thetop of the screen. This undoes the ERASE command and the 4 deleted items.

If you pick the Undo button again it will step back through the previous commands. For more advanced control, type the whole word **UNDO** at the command line. With this method you tell AutoCAD how many steps or command executions to go back. For example, the commands given below could be used to undo the last three operations.

Command: **UNDO**
Auto/Back/Control/End/Group/Mark/<Number>: **3**

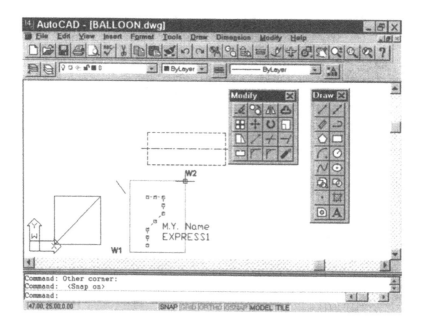

Figure 4.7 Erasing Window

HAZARD WARNING! This is a dangerous command as it can undo everything that you have done in the current editing session. If you then discover that you have undone too many commands the situation can be retrieved by executing a **REDO** immediately. The REDO button is beside Undo on the toolbar.

Command: **REDO**

Note that this command is more versatile than the "undo" which is available within the LINE command sequence. The general UNDO command cannot be executed transparently.

One of the best features of UNDO is creating marks. Before trying out some complicated manoeuvre you can set a mark. If things don't work out as planned you can use UNDO followed by Back to get back to the drawing as it was when you made the mark.

Command: UNDO
Auto/Back/Control/End/Group/Mark/<Number>: M
. . .
Do a series of AutoCAD commands.
Command: UNDO

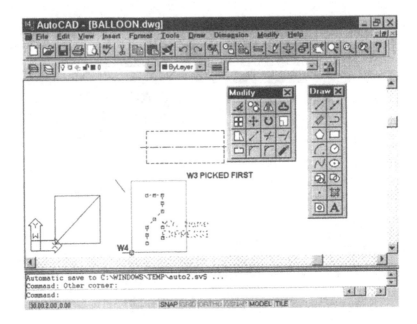

Figure 4.8 Erase Crossing

Auto/Back/Control/End/Group/Mark/<Number>: B

Note for Release 12 users: The OOPS command still works but does not appear on any of the usual toolbars. There is a button for it if you wish to customize a toolbar. There is little call for the OOPS command now that the Undo button is available.

Crossing windows and polygons

Now for something incredibly subtle. As mentioned above, the relative positions of the two corners of the window affect the behavior. If the first corner is to the right of the second, the "Crossing" method is used. For example to re-erase the previous lines, pick the point W3 **(47,25)** and then drag the other corner to W4, **(30,2)**, Figure 4.8. This selects the four lines as before but adds in the two lines of text. Now, pick the **ERASE** button and all six objects will disappear.

For the next couple of deletions we will use the Verb/Noun method. There is no particular significance in this choice. The selection methods used above

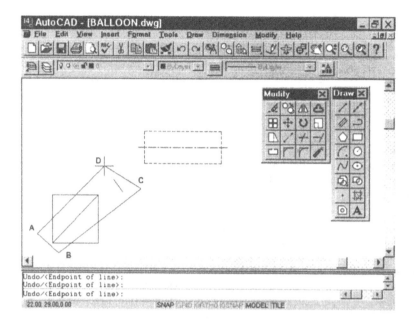

Figure 4.9 Erase Polygon

can equally be employed after picking ERASE as before. Likewise, the selection methods used below could be used with Noun/Verb procedures.

Firstly, to get rid of the lines indicated in Figure 4.9, the regular window is no good. The standard window is always rectangular and parallel to the screen axes. What is needed here is a skew window or "WPolygon". Pick the **ERASE** button or type the command.

Command: **ERASE**
Select objects: **WPOLYGON** (or just WP)
First polygon point: pick point **A**
Undo <Endpoint of line>: pick point **B**
Undo <Endpoint of line>: pick point **C**
Undo <Endpoint of line>: pick point **D**
Undo <Endpoint of line>: pick point **<ENTER>**
2 found.
Select objects: **<ENTER>**

This polygon works like the standard window. Only objects that are fully within the polygon are selected. The polygon can be any shape or size but its own lines must not cross. To tell AutoCAD that all the corners have been in-

Table 4.1 Selection options

Selection Keyword	Desctription
AUto	The default mode
Add	Switches back from Remove
ALL	Every object on thawed layers
BOX	Equivalent to Window
Crossing	Picks objects in and crossing a window
CPolygon	Picks objects in and crossing a polygon
Fence	Picks objects crossing line segments
Group	Objects within a specified group
Last	Most recently drawn object
Multiple	Does not highlight until end
Previous	The previous selection set
Remove	Removes objects from selection
SIngle	Picks one item and exits object selection mode
Undo	Cancels the most recent selection
Window	Picks objects lying completely within a rectangle
WPolygon	Picks objects lying completely within a polygon

put press <ENTER> without picking a point. With the Verb/Noun procedure, ERASE prompts you to select the objects. When the window has been completed AutoCAD echos the number of objects found and ghosts them. It then prompts you to select more objects. When you are done, just press <ENTER> and all the ghosted objects will go.

When using commands which prompt you to select objects, you can adopt any of the methods shown in Table 4.1. The uppercase letters indicate the permitted truncations. For some reason, the standard AutoCAD menus do not explicitly list these. It is worth keeping a note of these and maybe customizing the AutoCAD menus to include the most useful ones.

Table 4.1 is based on the Help file information for the SELECT command. For the noun verb operations earlier you could only select objects by single picks or implied windows. To do more complicated selections and to build up the selection set, type **SELECT** at the command line. The SELECT command is the explicit form for creating a selection set of objects. It allows you to use any of the methods in Table 4.1. Again, you select the objects before deciding what to do with them. Here, we will use the "Fence" option to select four lines ghosted in Figure 4.10 and then delete them. The fence selects all objects that it crosses.

Command: **SELECT**

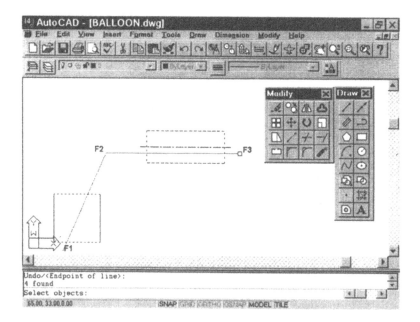

Figure 4.10 Fence selection method

Select objects: **Fence**
First fence point: pick point **F1**
Undo/<Endpoint of line>: pick point **F2**
Undo/<Endpoint of line>: pick point **F3**
Undo/<Endpoint of line>: **<ENTER>**
4 found
Select objects: **<ENTER>**

To delete these four lines you need to execute the ERASE command and call up the "Previous" selection set, ie the one just done.

Command: **ERASE**
Select objects: **Previous**
4 found
Select objects: **<ENTER>**

The selection process used with the SELECT and ERASE commands is exactly the same for many other commands. Each object that has been ghosted becomes part of a "selection set". The selection set can have more objects added to it or have some removed.

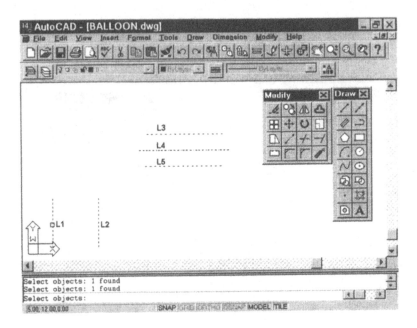

Figure 4.11 Erasing the last lines

Of the other options in Table 4.1, Last, ALL, CPolygon, Add and Remove are the most useful. "ALL" selects everything in the drawing excluding objects on invisible and frozen layers. This is usually done in conjunction with the "Remove" option, eg select everything except... This is good for stripping away unwanted detail. Typing or picking the keyword "Remove" changes the selection process from include to exclude mode. After picking "Remove" any objects selected are taken out of the selection st . "Add" goes back to the "include" mode. Try this out with the last five lines shown in Figure 4.11.

Command: **ERASE**
Select objects: pick lines **L1** to **L5**
Select objects: **Remove**
Remove objects: pick line **L1** 1 found, 1 removed.
Remove objects: **<ENTER>**

That should have un-ghosted line L1 and saved it from deletion. However, the stay of execution is only temporary. As we need a clean sheet for the rest of the exercise delete it as follows.

Command: **ERASE**
Select objects: pick line **L1** 1 found.

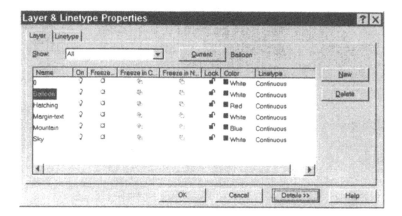

Figure 4.12 Layers for Balloon drawing

Select objects: **<ENTER>**

The Balloon drawing environment

The drawing limits and linetypes etc. have been inherited from the file, EX-PRESS1.dwg. That drawing contained only one layer, namely the default 0 layer. In the following part of the exercise you will need an additional five layers with appropriate color settings. To set up the layers shown in Figure 4.12 pick **Settings/Layer Control** ... from the menu bar. Then move the cursor arrow to the input box and type the layer names, "BALLOON, HATCHING, MARGIN-TEXT, MOUNTAIN,SKY". Then pick the **New** button.

Set the HATCHING layer color to red and Mountain to blue by double clicking on the color box or the word "White" on the respective lines. In the color pop up pick the colors from the standard list at the top. Then set Balloon to be the current layer and pick **OK**. If you have trouble doing this refer back to Figure 3.3 on page 42.

Creating circles, arcs and ellipses

Circles

Circles can be drawn in a variety of ways. The first circle will be specified by its centre point and radius while the second by centre and diameter. Pick

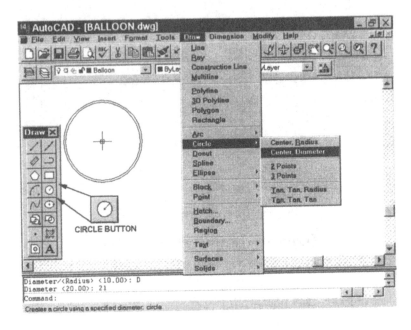

Figure 4.13 Concentric circles

the **Circle button** from the tool box. If the Draw toolbar is not visible use View/Toolbars to get it. Note that if the Draw toolbar is docked at one side of the screen the toolbar title will not show. You may have to search the screen to find the Circle button.

Command: **CIRCLE** 3P/2P/TTR/<Center point>: **20,30**
Diameter/<Radius>: **10**
Command: **<ENTER>**
CIRCLE 3P/2P/TTR/<Center point>: **@**

This is equivalent to @0,0 from the last point, ie the centre of the first circle. Type **D** or **Diameter** to change the prompt.

Diameter/<Radius>: **Diameter**
Diameter <20.0000>: **21**
Command:

The second circle could have been input with a radius of 10.5. We now have two concentric circles (Figure 4.13). On some displays it might be difficult to distinguish between the two circles. Use ZOOM to check that they are really there.

All the methods of circle creation appear in the **Draw/Circle** pull-down menu. The two point circle is identified by picking the ends of one of its diameters. The three point circle allows some nice geometric constructions. The TTR allows you to draw a circle tangential to two objects and with an input radius. More circles will be covered in Chapters 7 and 8.

Many ways to draw an ARC

Before adding a cloud to the sky we must switch layers. The easiest way to do this is to pick the layer name box on the Object Properties toolbar. This displays a pull-down list of all the layers. There is a scroll bar to the right of this pull-down for moving up and down long lists of layers. Pick **SKY** from the list and the current layer should change (Figure 4.14).

If you pick **Draw/Arc** from the menu bar you get a long list of arc creating options. At first this is a bit daunting with all the similar looking methods. There is also an Arc button in the tool-box.

To draw a cloud for the sky beside the balloon, you will need six arcs. The first arc will be a semi-circle. Since the angle within a semi-circle is 180 degrees the **Start End,Angle** (Figure 4.14) option is an appropriate choice. If you had to alter your units in Chapter 2, you will have discovered that angles are measured positive in the anti-clockwise (or counter-clockwise) direction. This means that the start and end points should always be picked so that the arc joining them will go anti- clockwise.

Command: **ARC** Center/<Start point>: **55,33**   (A1)
End point: **47,33**   (A2)
Angle/Direction/Radius/<Center point>: A Included angle: **180**

The second arc can be drawn using the **Start,End,Radius** with a radius of 3.

Command: **ARC** Center/<Start point>: **48,34**   (Arc B)
End point: **@5<215**
Angle/Direction/Radius/<Center point>: R Radius: **3**

Do the next two arcs with the **Start,Centre,End** approach.

Command: **ARC** Center/<Start point>: **45,32**   (Arc C)
Center/End/<Second point>: C Center: **45,29**
Angle/Length of chord/<End point>: DRAG **@5<326**
Command: **ARC** Center/<Start point>: **45,27**   (D1)
Center/End/<Second point>: C Center: **50,32**   (D2)
Angle/Length of chord/<End point>: DRAG **@11<307**   (D3)

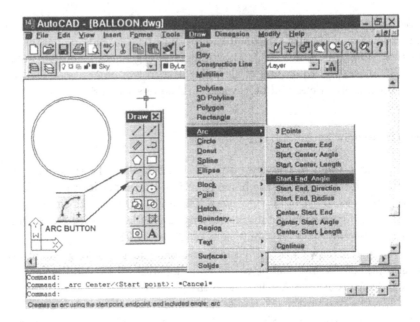

Figure 4.14 Arc menus

The angles used for the end points look a bit strange, but they are the result of dragging the arc until it looked right. You can use your judgement to get the arcs looking like Figure 4.15. The actual arc end points are not on the points C3 and D3, but are on the intersection between the arc whose radius is calculated from the start and centre points, and the line from the centre to the points C3 and D3. Figure 4.15 shows this procedure for arc D.

The remaining two curves to finish off the cloud can be drawn with **3 Point** arcs.

    Command: **ARC** Center/<Start point>: **52,26**
    Center/End/<Second point>: **57,25**
    End point: DRAG **59,32**
    Command: **ARC** Center/<Start point>: **59,30**
    Center/End/<Second point>: **60,35**
    End point: DRAG **54,34**

Rectangles and ellipses

You can make rectangles by drawing four connected LINEs, but this would be an assembly rather than a single entity. Thus to erase the rectangle formed

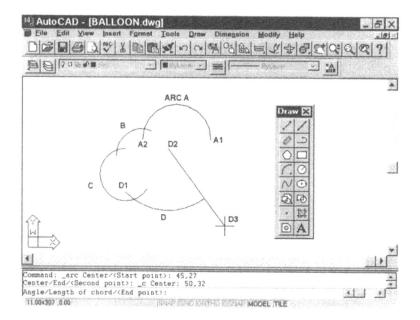

Figure 4.15 Dragging the ARC

with LINEs you would have to pick all four sides in the selection of objects. The difference with polylines is that all the segments are considered as part of one entity. PLINE can be used in all the ways that LINE is used and a few more besides. Firstly, you can use it in place of LINE to draw the rectangle for the balloon's basket (Figure 4.16). Some special shapes use polylines in their construction eg rectangles, donuts and polygons.

Change back to layer **BALLOON** and draw the rectangular gondola (ABCD in Figure 4.16) beneath the balloon by picking **Draw/Rectangle** or picking the Rectangle button on the Draw toolbar. This constructs a rectangle using a polyline. For example, the ABCD rectangle could have been drawn as follows:

Command: _rectang

Chamfer/Elevation/.../<First corner>: **15,10**   (A)

Other corner: **@10,5**   (C)

If your rectangle appears with very thick lines then your polyline width has probably got a non-zero value. Skip forward to the section on Wide Lines on page 95 and set this to zero, erase the thick rectangle and draw it again.

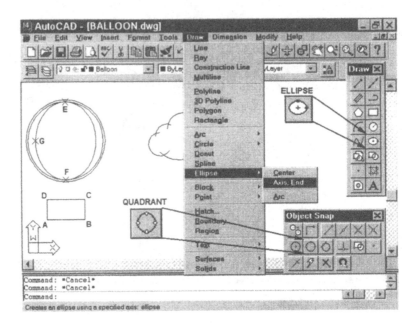

Figure 4.16 Basket and ellipse

Note that this polyline rectangle is one single entity. If you try to **ERASE** it just pick one point on the rectangle and the whole thing will be selected. Use ^C to avoid rubbing it out or **OOPS** to bring it back.

To draw the ellipses on the smaller circle of the balloon pick **Draw/Ellipse** and **Axis End** from the menu bar or use the Ellipse button on the Draw toolbar. You will be prompted to give the two end points of one of the axes and one end of the other axis. In this example, you will use object snap QUADrant to locate the top and bottom of the inner circle for the major axis. Make sure that the Object Snap toolbar is available; pick **View/Toolbars...** and select Object Snap.

Command: **ELLIPSE**
<Axis endpoint 1>/Center: **QUADRANT of pick point E**
Axis endpoint 2: **QUADRANT of pick point F**
<Other axis distance>/Rotation: **12,30**   (G)

Repeat this procedure for two more ellipses to complete the balloon (see Figure 4.17 below).

Command: **ELLIPSE**

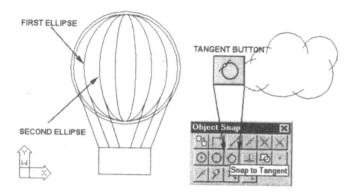

Figure 4.17 More ellipses and cables

<Axis endpoint 1>/Center: **QUADRANT** of **pick point E**
Axis endpoint 2: **QUADRANT** of **pick point F**
<Other axis distance>/Rotation: **15,30**
Command: **<ENTER>**
ELLIPSE
<Axis endpoint 1>/Center: **QUADRANT** of **pick point E** (20,40)
Axis endpoint 2: **QUADRANT** of **pick point F** (20,20)
<Other axis distance>/Rotation: **18,30**

To connect the basket to the balloon, some cables must be provided. Use SNAP to locate the end points on the basket and object snap TANGENT on the ellipses. It may help to use Zoom to get a closer look at the balloon when picking the inner circle. Pick **Draw/Line** from the menu bar.

Command: **LINE** From point: **15,15**   (D)
To point: **TANGENT** To **10,30**   (inner circle)
To point: **<ENTER>**
Command: **<ENTER>**
LINE From point: **17,15**
To point: **TANGENT** To pick the first ellipse on the left

Make sure that you pick the correct ellipse i.e. near the point (12,30). Now try the second ellipse.

To point: **<ENTER>**
Command: **<ENTER>**
LINE From point: **19,15**

        To point: **TANGENT** To **16,30**   (second ellipse)
        To point: **<ENTER>**

And for the cables on the right-hand side:

        Command: **<ENTER>**
        LINE From point: **21,15**
        To point: **TANGENT** To **25,30**   (second ellipse)
        To point: **<ENTER>**
        Command: **<ENTER>**
        LINE From point: **23,15**
        To point: **TANGENT** To **28,30**   (first ellipse)
        To point: **<ENTER>**
        Command: **<ENTER>**
        LINE From point: **25,15**
        To point: **TANGENT** To **30,30**   (inner circle)
        To point: **<ENTER>**

Adding text

Textual annotations are an important part of any engineering drawing. The annotations may include information about the objects on the drawing or about the drawing itself. AutoCAD allows you to insert text on a drawing giving you full control over how this should be done and how the text should look.

Already in the drawing EXPRESS1 you have added your name using the DTEXT command. In that example you picked the start point of the text and its height and rotation. To repeat a similar operation, switch to the layer MARGIN-TEXT and try the following. You can switch layers by picking the box with the layer name, Balloon, from the Object Properties tool bar and clicking **Margin-text**.

There are two different types of text in Release 14. The basic text entity is created either by the DTEXT command or the aptly named TEXT command. The DTEXT and TEXT behave almost identically with DTEXT being a bit more versatile. The TEXT command is a hangover from earlier versions of AutoCAD. With Release 14 a new text entity called multiline or MTEXT is available. Here we will input one line of text using DTEXT and then a couple using MTEXT for comparison. The text button, denoted by a large letter "A" on the Draw toolbar runs MTEXT. To use DTEXT you must type the command or pick **Draw/Text/Single Line Text**.

        Command: **DTEXT** or pick the button

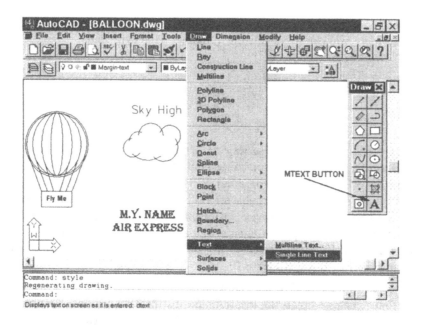

Figure 4.18 Adding Text

Justify/Style <Start point>: **45,40**
Height <2.00>: **2.5**
Rotation angle <0>: **ENTER>**
Text: **Sky high <ENTER>**
Text: **<ENTER>**

Despite the pull down menu saying "Single Line Text" (Figure 4.18), DTEXT expects multiple lines of text. This gives left justified text, i.e. the first letter is lined up with the start point. Other types of justification are allowed by typing **J** at the "Justify/Style <Start point>" prompt. It can be used to centre the characters on a point or to right justify the text.

The most common mistake people make when entering text is not paying attention to the prompt line. As most text is inserted with an angle of 0 degrees it is easy to forget that you have to re-input the angle every time. This type of error is more likely to crop up if you are inputting a lot of text and paying more attention to the difficult operation of typing than to the screen.

To make the most impact with text, it must look good. The default type of lettering available is called the SIMPLEX font (on some systems TXT is the default) which is simple, fast and memory efficient. AutoCAD provides a number of different text fonts ranging from the simplicity of SIMPLEX to the extremely complicated and flamboyant GOTHIC. In addition to the AutoCAD provided fonts you can access all the Windows TrueType fonts.

Text styles

Multiline text gives the easiest way to utilise Truetype fonts. It also gives easy to use facilities for incorporating many lines of text (which can be imported from other Windows programs). With much better editing capabilities, MTEXT is an effective replacement for DTEXT and should cater for virtually all of your needs. The only proviso with using MTEXT is that if you regularly send drawings to colleagues or clients who are using earlier versions of Auto-CAD then MTEXT entities need to be converted to basic text entities when exporting the file to, say, Release 12.

Before inputting the Name and Title for the drawing in an appropriately arcaic font it helps to plan the location and justification options. Keeping the text height at 2.5 you can calculate the size of the text box required to contain all the characters. If in doubt, make the box bigger than you need. A box about 30 units wide is sufficient. Pick the MTEXT icon on the Draw toolbar or use **Draw/Text/Multiline text** from the menu bar.

> Command: _mtext Current text style: Standard. Text height: 2.5
>
> Specify first corner: **35,10**
>
> Specify opposite corner or [Height/Justify...]: **@30,-8**

This specifies an area of the drawing for the text. If the area turns out inappopriate, it can be altered later. The Height, Justification etc. can all be specified in the resulting dialog box, Figure 4.19.

Select the Truetype font **Algerian** from the pull down list. A quick way to get this is to type the first letter of the font name when the pull down list is shown. This snaps to the appropriate part of the alphabetical list. Now pick the Properties tab and pick **Top Center** from the justification list. This centers the text making the first line flush with the top of the invisible text box specified at the start of the command sequence. Other options allow centering about the middle of the box or the bottom or using left or right justified text.

Now press the mouse button when the cursor is within the text display area and start typing: **M.Y. NAME AIR EXPRESS**. Note that the word wrapping is automatic and depends on the width of the invisible text box. In order to get the correct line break position the cursor just before the word "Air" and press <**ENTER**>. Finally, pick the **OK** button.

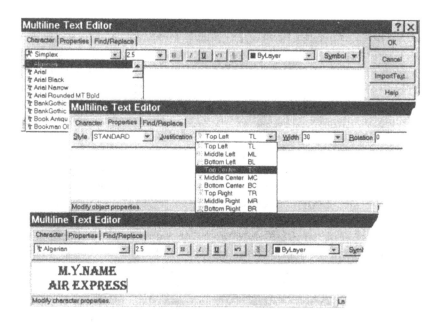

Figure 4.19 Mtext editor

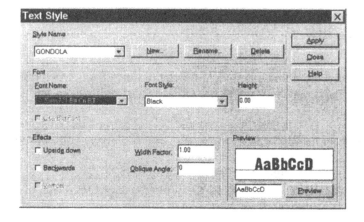

Figure 4.20 Text style Gondola

Style command

AutoCAD has a rather confusing arrangement of text fonts and styles. The best way to avoid confusion is to select fonts using Mtext as outlined above. Despite this author's confusion with AutoCAD's text styles it is often advantageous to create drawings with a number of text styles.

You may have noted that both with DTEXT and MTEXT there was the opportunity to alter the "STYLE" of the text. In the examples above only the "STANDARD" style was used. This was the AutoCAD default.

Since many drawing offices will adopt house styles and because you will sometimes need to create your own style here is how it's done. You can create a new text style by picking **Format/Text Style...** from the menu bar. This brings up the Text Style dialog box (Figure ref{f420). Pick **New** and name the new style **Gondola** and pick **OK**. Now pick the font as **TT Swiss** or similar from the font name pull down menu. If you do not have this font pick an alternative from your list. Note that the Font Style is "black" and the height is zero. The zero height is important as it implies that the user will be prompted for the height each time text is added. If a non zero height is given in the text style then that height is always used. This is usually a mistake. Always leave the height as zero unless you have a specific need for a constant height. Finally, pick **Apply** button to set the new style and then pick **Close** to get rid of the dialog box. All new text will now be drawn in the new style, Gondola.

To put the final bit of text within the gondola beneath the balloon use Dtext's justified option.

> Command: **DTEXT** or pick **Draw/Text/Single Line Text**
> Justify/Style <Start point>: **J**
> Align/Fit/Center/...: **C**

You must use the abbreviated form for the option. If you type "center", AutoCAD will assume you want to use Object Snap and tries to find the centre of an arc or circle.

> Center point: **20,12**   (middle of basket)
> Height <2.5>: **1.8**
> Rotation angle <0>: **<ENTER>**
> Text: **Fly Me**
> Text: **<ENTER>**

There are lots of the other justifying options: inserting text using the "Middle" option is similar to centred but text is balanced about the point horizontally and vertically. With "C" the text is balanced horizontally. With "Aligned" text you give the start and end points and AutoCAD calculates the height so that your text just fits. The rotation angle is defined by the two end points. "Fit" is similar to aligned but you can specify a text height and AutoCAD calculates the text width. The final option, "Right", allows you to input the end point, height and rotation of the text and AutoCAD works out the start point. Finally, the abbreviations, TL ... BR indicate the top left or bottom right corners of the text. These last ones are the same as the options for Mtext.

You can change the text style by typing "S" as the reply to the "Justify/Style..." prompt. This will only show the styles that have been defined in the current drawing. When using Mtext you can use the Style pull down menu which appear when the Properties tab is selected.

On serious drawings it is important to keep control of the text heights that are used. Too many different heights and/or fonts spoil the appearance of the drawing. Furthermore, it is vital that all text be legible on the final plot or print out. Thus, the text heights should be chosen carefully and with a view to the final legibility.

On large drawings the time taken to regenerate all the text can be considerable. While editing, you can speed things up by using the simple fonts, TXT and SIMPLEX, and changing to the fancy ones at the end. You can also put all the text on suitable layers and freeze them until the end. The disadvantage of the latter approach is that while the text is invisible, you may draw something on top of it. An alternative approach is to turn on the QTEXT or quick text mode. When QTEXT is on, all the text is displayed as rectangles. With this you at least know where the text is. **Quick Text** is on the Drawing Aids dialog box and can be found by picking **Tools/Drawing Aids**... from the menu bar. Then click the box beside "Quick Text" followed by **OK**. The text display will not change until the next REGEN command is issued. You can also do this by typing the command.

Command: **QTEXT** ON/OFF <OFF>: **ON**
Command: **REGEN**   (to see the effect)

Turn **QTEXT** back **OFF** and draw the margin ABCD shown in Figure 4.21.

Command: **QTEXT**
ON/OFF <ON>: **OFF**
Command: **REGEN**

Wide lines

Up to this point, we haven't been too concerned with the width of the lines we were drawing. In fact everything we have drawn has had a notional line width of zero. AutoCAD uses zero width to mean that no matter how much

Figure 4.21 Quick text mode in the frame

you magnify the line it always appears the same width. Similarly, it doesn't matter how far you zoom out from an object, the width doesn't change.

There are many times when you might want to draw a line with a specified width: e.g. printed circuit board drawings. In this section you will cover the methods of assigning widths and using them in the balloon drawing artwork. The same commands and methods can be used in any type of drawing where wide lines are required.

WARNING! Assigning width is mildly dangerous for plotted output. Assigning non-zero widths should only be used where the width serves a particular purpose. If you have a drawing convention that, say, all outlines are to be drawn in 0.5mm line width then it might be better to assign a particular color to the layer containing the outlines. Then, at plot time, you can specify that the outline color will be drawn using a 0.5mm pen. If you assign a width on the AutoCAD drawing itself and plot the drawing at a scale of 1:10 the width will be plotted at the reduced width of 10% the original.

The best way of generating wide lines is to use the Polyline or PLINE command. With polylines it is possible to assign a constant width or a varying width to the line. You can also draw wide arcs with PLINE. To draw a frame for the balloon drawing (Figure 4.21) use **Draw/Polyline/2D** from the menu bar and set a width of 0.5 units. Make sure ORTHO is on and SNAP is set to 1 and that the current layer is MARGIN-TEXT.

```
Command: PLINE
From point: 1,1                                              (A)
Current line-width is 0.0000
Arc/Close/.../Width/<Endpoint of line>: Width
```

Starting width <0.0000>: 0.5
Ending width <0.5000>: <ENTER>
Arc/Close/.../<Endpoint of line>: **64,1** (B)
Arc/Close/.../Width/<Endpoint of line>: **64,44** (C)
Arc/Close/.../Width/<Endpoint of line>: **1,44** (D)
Arc/Close/.../Width/<Endpoint of line>: **Close** (A)

If your polyline is drawn in outline only, then you should turn the **Solid Fill** on from the Drawing Aids dialog box (**Tools/Drawing Aids...**) and REGENerate the drawing or type the command.

Command: **FILL**
ON/OFF <current>: **ON**
Command: **REGEN**

Filling in wide lines and solid objects can slow down AutoCAD's responses. Turning FILL off will speed up the REDRAW and REGEN commands on large drawings.

In the following sequence of operations you will add some mountain scenery to the picture (Figure 4.22). Before putting in the Blue Ridge Mountains we must switch to the layer, MOUNTAIN. Click the Layer name box which currently reads "Margin-text" from the Object Properties tool-bar and then click **MOUNTAIN** from the resulting list. The color for this layer was previously set to blue.

The first PLINE will have a series of segments with two different but constant widths to generate the mountains on the right of Figure 4.22. The second will have varying widths to make the slope on the left of the picture. The listing of the commands and prompts looks long, but most of the work is done by AutoCAD. If you pick any wrong points or get the widths mixed up, use the "Undo" option to step backwards along the polyline.

Command: **PLINE**
From point: **29,4** (F)
Current line-width is 0.5000

Note that the width is retained from the last command. Change it back to zero.

Arc/Close/.../Width/<Endpoint of line>: **W**
Starting width <0.50>: **0**
Ending width <0.00>: <ENTER>
Arc/Close/.../Width/<Endpoint of line>: **@6,10** (G)
Arc/Close/.../Width/<Endpoint of line>: **@4,−2** (H)
Arc/Close/.../Width/<Endpoint of line>: **@4,5** (J)

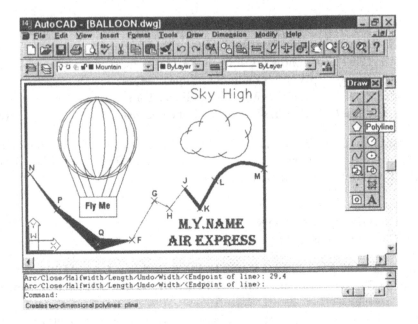

Figure 4.22 Varying PLINE's width

Now change the width to 0.75 units.

> Arc/Close/.../Width/<Endpoint of line>: **W**
> Starting width <0.00>: **0.75**
> Ending width <0.75>: **<ENTER>**
> Arc/Close/.../Width/<Endpoint of line>: **@4,−5**   (K)
> Arc/Close/.../Width/<Endpoint of line>: **@8<60**   (L)

Now change over to drawing a poly-arc.

> Arc/Close/.../Width/<Endpoint of line>: **Arc**
> Angle/CEnter/.../Width/<Endpoint of arc>: **@13,3**   (M)
> Angle/CEnter/.../Width/<Endpoint of arc>: **<ENTER>**

The arcs in a polyline are always drawn tangential to the previous PLINE segment. This ensures a smooth transition between the straight lines and the curve. When the command sequence has been completed, AutoCAD calculates the chamfering required for each corner.

The mountains on the other side of the balloon will be made up of varying width or tapering polylines. We will start with a zero width, increasing to 1 unit and then 3.

Command: **PLINE**
From point: **2,21** (N)
Current line-width is 0.75
Arc/Close/.../Width/<Endpoint of line>: **W**
Starting width <0.75>: **0**
Ending width <0.00>: **1**
Arc/Close/.../Width/<Endpoint of line>: **9,12** (P)
Arc/Close/.../Width/<Endpoint of line>: **W**
Starting width <1.00>: **<ENTER>**
Ending width <1.00>: **3**
Arc/Close/.../Width/<Endpoint of line>: **20,3** (Q)

And now to reduce the width back to zero and connect up with the first previous polyline mountainscape.

Arc/Close/.../Width/<Endpoint of line>: **W**
Starting width <3.00>: **<ENTER>**
Ending width <3.00>: **0**
Arc/Close/.../Width/<Endpoint of line>: **29,4** (F)
Arc/Close/.../Width/<Endpoint of line>: **<ENTER>**

The final polyline will be used to draw the silhouetted bird in flight (Figure 4.23). Every self-respecting sky should have one, so change to the **Sky layer** by picking the layer list from the Object properties toolbar. The bird is achieved by starting out with a straight line segment with a width going from zero to 0.4. This is then blended with an arc whose width increases from 0.4 to 0.8 followed by a short line segment. The reverse procedure gives the other wing. As this is an intricate manoeuvre, it is prudent to ZOOM in first.

Command: **ZOOM**
All/Center/.../Window/<Realtime>: **W**
First corner: **29,16**
Other corner: **42,24**
Command: **PLINE**
From point: **30,18** (A)
Current line-width is 0.00

Set up varying width.

Arc/Close/.../Width/<Endpoint of line>: **W**
Starting width <0.00>: **<ENTER>**
Ending width <0.00>: **0.4**
Arc/Close/.../Width/<Endpoint of line>: **@2,1** (B)
Arc/Close/.../Width/<Endpoint of line>: **W**

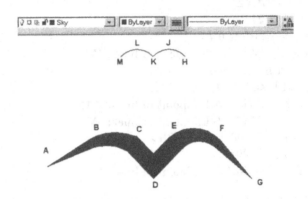

Figure 4.23 Silhouetted birds

Starting width <0.40>: **<ENTER>**
Ending width <0.40>: **0.8**

Change over to drawing arcs.

Arc/Close/.../Width/<Endpoint of line>: **ARC**
Angle/CEnter/.../Width/<Endpoint of arc>: **@2,0**   (C)

Change back to straight lines. Note that the chamfers will not be calculated
until the command sequence has been completed.

Angle/CEnter/.../Width/<Endpoint of arc>: **Line**
Arc/Close/.../Width/<Endpoint of line>: **@1,−1**   (D)
Arc/Close/.../Width/<Endpoint of line>: **@1,1**   (E)

And draw the second half of the bird.

Arc/Close/.../Width/<Endpoint of line>: **W**
Starting half-width <0.80>: **<ENTER>**
Ending half-width <0.80>: **0.4**
Arc/Close/.../Width/<Endpoint of line>: **ARC**
Angle/CEnter/.../Width/<Endpoint of arc>: **@2,0**   (F)
Angle/CEnter/.../Width/<Endpoint of arc>: **Line**
Arc/Close/.../Width/<Endpoint of line>: **W**
Starting width <0.40>: **<ENTER>**
Ending width <0.40>: **0**

For the final wingtip, use the **Length** option. This adds a line of the specified length, tangential to the last arc.

> Arc/Close/.../Length/Undo/Width/<Endpoint of line>: **L**
> Length of line: **2.24**   (G)
> Arc/Close/.../Width/<Endpoint of line>: **<ENTER>**

Note that the subtle differences in the construction of each wing yields an unsymmetric result. This is principally because when finishing the first half of the bird, a relative coordinate was used rather than specifying the length option. Thus, that line is not tangential to the arc. On the second part of the bird, all lines are tangential to the arc.

A second bird is flying in the distance. It can be made with just two ordinary three-point ARCs. Pick **Draw/Arc/3 Points** from the menu bar.

> Command: **ARC** Center/<Start point>: **36.5,23**   (H)
> Center/End/<Second point>: **35.5,23.3**   (J)
> End point: DRAG **35,23**   (K)
> Command: **ARC**
> Center/<Start point>: **@**   (K)
> Center/End/<Second point>: **34.5,23.3**   (L)
> End point: DRAG **33.5,23**   (M)

Aerial view

It can sometimes become a bit annoying to have to zoom in and out on a drawing repeatedly to locate the next bit of drafting. This is particularly so on large drawings. AutoCAD for Windows provides a facility called "Aerial View" which helps to navigate through drawings without repeated zooms. The Aerial View button looks like a miniature airplane and is located towards the right of the Standard toolbar (Figure 4.24).

Click this button and the Aerial View Window appears. The **Zoom** magnifying glass button should be highlighted. The window shows the complete drawing with a box indicating the current zoom window. To select a new zoom window pick the lower left corner and then the other corner, within the Aerial View window. The view in the main window, behind the aerial view should update automatically. If it does not, pick **Options/Dynamic Update** from the Aerial View menu bar.

One of the most useful features of Aerial View is the ease with which one can pan around the drawing. Pick the Pan icon in the Aerial View window to switch to pan mode. Then drag the zoom frame around the window and pick the button. The background main window will update when the mouse button is pressed. This is good for getting around large drawing.

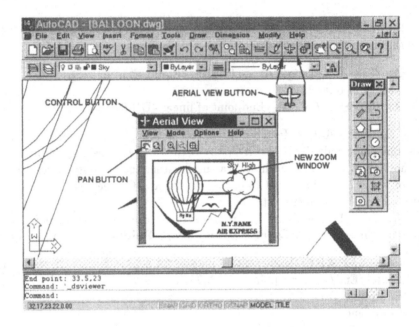

Figure 4.24 Aerial view of balloon

Make sure that you end up with a zoom window as shown in Figure 4.24 so that you can draw the sun in the space between the balloon and the "Sky high" text.

You can resize or move the Aerial View window around the screen like in any other Windows program. You can move it out of the way by picking the title bar and dragging it around. To get back to the drawing screen, just move the cursor to the drawing area and click. The Aerial View can be closed by picking control box in the top left corner followed by Close.

Donuts and discs

Here's one final polyline object, a filled circle. The amusingly named DONUT (can also be spelt "DOUGHNUT") is in fact a little program based on the PLINE Arc. It can be found on the DRAW screen menu. As the name suggests you can use it to draw fat circles. If the inside diameter of the doughnut is zero then you have a filled circle. To draw the sun in in the sky (Figure 4.28) you will first have make sure that you are working on the Sky layer and change the entity color from bylayer to yellow.

Command: **COLOR**   (Or pick the color box from the tool-bar)

New entity color <BYLAYER>: **YELLOW**

Draw the sun. Pick **Draw/Donut** from the menu bar.

Command: **DONUT**
Inside diameter <0.50>: **0**
Outside diameter <1.00>: **6**
Center of doughnut: **37,40**
Center of doughnut: **<ENTER>**
Command: **COLOR**
New entity color <2 (yellow)>: **BYLAYER**
Command: **ZOOM**
All/Center/.../<Scale(X/XP)>: **A**

This means that the colors of new entities will be the same as their layer's default color. This was set up using the LAYER command.

Shading with patterns

The pervious sections used polylines to generate 2D representations of solid or filled shapes. The chief difficulty with using PLINE for creating, say, a solid square is that you have to draw the square as a polyline along the centerline and specify the width equal to that of the square. It is often more sensible to draw the outline of the square and shade it with some suitable pattern.

Many types of architectural and engineering drawings use standard hatching patterns to indicate such things as cross sections and material type. AutoCAD contains a good library of hatch patterns conforming to ANSI (American National Standards Institute) norms. There is also a hatch pattern for creating a solid fill.

There are two main methods of shading objects in AutoCAD. Firstly, the Boundary Hatch (or BHATCH command is a versatile facility for shading complex areas, with ot without holes. Secondly, AutoCAD allows you to apply Postscript shading patterns to closed 2D polyline shapes. This is less versatile but is very good for representing solids in 2D drawings. PSFILL is good for presentation graphics and for those users with experience of postscript.

For BHATCH to work correctly, the area to be shaded should be within a closed boundary. Furthermore, the entities, lines, arcs etc, making up the perimeter must intersect at their end points. If they don't, or if there are any protruding ends, the results may be incorrect with some strange shading.

It is a good policy to make a small test box for hatching so that the effectiveness of the pattern scale can be assessed before hatching a larger area.

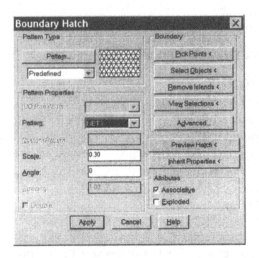

Figure 4.25 Boundary Hatch dialog box

Command: −**LAYER**   or pick the pull-down layer list
?/Make/Set/...: **S**
New current layer <0>: **HATCHING**
?/Make/Set/...: **<ENTER>**
Command: **LINE**
From point: **33,9**
To point: **@5,0**
To point: **@0,−5**
To point: **@−5,0**
To point: **Close**
Command: **SAVE**
File name <BALLOON>: **<ENTER>**

HAZARD WARNING! The BHATCH command can be dangerous. Always
save your drawing before executing this command. BHATCH has some safety
features

The procedure for adding patterns to a drawing is controlled through
the command **BHATCH** or Boundary HATCH. It is executed by picking
Draw/Hatch... from the menu bar.

Command: **BHATCH**

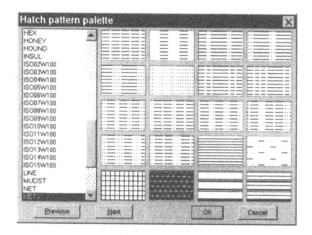

Figure 4.26 Hatch pattern palette

This brings up the Boundary Hatch dialog box shown in Figure 4.25. You will then have to select a hatch pattern followed by picking the object defining the boundary of the hatch area.

To select the hatch pattern pick the **Pattern** button. Make sure that the Pattern Type, "Predefined" is active. This appears in a pull down list just below the "Pattern" button. If the field displays "User- Defined" or "Custom" simply pick the box and change the setting. The Hatch Pattern dialog box should now appear (Figure 4.26).

Scroll down the list of pattern names headed by "Solid". Alternatively you can pick the **Next** button until the NET3 pattern appears. A quick way to move through the list is to highlight any pattern name. Then type N. This zaps you to the first pattern whose name starts with N. NET3 is just below this. Having highlighted **NET3**, pick **OK**.

This brings you back to the Boundary Hatch dialog box with the correct pattern, as shown in Figure 4.25. One last thing to do here is to select a scale for the pattern. As our drawing limits are relatively small, the default scale 1.00 will have to be reduced. Move the cursor into the Scale box and type **0.30**. Then click **OK**.

Back at the Boundary Hatch dialog box select the **Pick Points** button in the top right. Pick any point within the red square or type the coordinate **35,6**.

Select internal point: **35,6**   (A point within the red square)
Selecting everything...
Selecting everything visible...

Analyzing the selected data...
Analyzing internal islands...
Select internal point: **\<ENTER\>**

This method of selection is suitable for relatively simple shapes. AutoCAD analyzes the drawing to identify the boundary entities. If you pick a complicated shape or mistakenly pick outside the square, this analysis could take a long time. You can abort by pressing the **Esc** key. If you pick the **View Selections** you can see which objects have been selected.

Before finally executing the command pick **Preview Hatch**. This will show how the shading will look. Press **\<ENTER\>** or pick **Continue** to go back to the dialog box. If the pattern seemed satisfactory you can proceed to shade the box by picking **Apply**. Only on picking "Apply" is the command executed. If the hatching is not as shown in Figure 4.28, then re-check pattern and scale. If you have already applied the hatch and it looks wrong, ERASE it and try again. Make sure that you use the correct scale and that the four sides of the square are selected.

Now to shade in the stripes on the balloon. As we will be using the same pattern and settings as before we only need to select the object, preview and apply the hatch from the Boundary Hatch dialog box. Pick **Draw/Hatch**....

You might have noticed that one of the things AutoCAD did in the last sequence was to "remove islands". If the square had had circle or some other closed shape inside it this would have been an island. AutoCAD gives a few options on how to deal with islands.

Pick the **Advanced**... button. The Advanced Options dialog box appears (Figure 4.27). Then pick **Normal** from the Boundary style and pick **OK**. To select the stripes use a window and include the inner circle and all the ellipses of the balloon but exclude the outer circle. The Normal method is to alternate the hatching on and off for each successive boundary. To do this for the balloon stripes check that the Hatching Style is set to "Normal". This means that if there are inner boundaries then the hatch will be applied to alternate areas starting with the outermost.

Of the other options for hatching style, "Outer" shades in only the outermost region. "Ignore" simply ignores any internal boundaries. To see how these affect the circle, square and triangle on the dialog box you can pick the appropriate buttons. Make sure that you reset it to "Normal" before proceeding.

Then pick **Draw/Hatch**....

Command: _bhatch
Select objects: **W**
First corner: **10,20**
Other corner: **30,40**

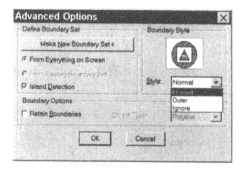

Figure 4.27 The AutoCAD Air Express

4 found.
Select objects. **<ENTER>**

The shading shown in Figure 4.28 results because the ellipse boundaries have been selected within other boundaries. If you are happy with the hatching then press **<ENTER>** to go back to the dialog box and pick **Apply** to execute the command.

Note that selecting the objects was necessary in this example. If you had used "Pick points" you would have had to pick each region within the balloon including the parts bounded by the cables to the gondola. Selecting the object with a window was far simpler.

Despite the hazard warning, there is little to fear from shading in objects with patterns. Take heed of the warning and follow the advice on saving the drawing. The reason for this caution is that if the pattern scale is small then it will use up a lot of memory. Hitting the **Esc** will stop a rogue hatch in progress.

As with text, the more complicated the pattern the more it will slow down your redraw time. You will save time if you can leave any hatching as late as possible in the drawing. Freeze the layer when the hatching is not required to be displayed.

To finish this session, pick **File/Exit** followed by picking **Save Changes**.

Summary

You have encountered AutoCAD's most important drawing commands and entity types in this chapter. You have also used AutoCAD's entity selection procedure. This is common to many commands incuding **ERASE** and **HATCH**. Polylines are versatile entities combining lines, arcs, traces and solids. You can

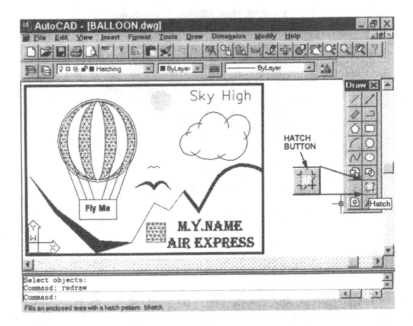

Figure 4.28 The AutoCAD Air Express

speed up REDRAWs by turning FILL off, QTEXT on and freezing the hatch layers.

You should now be able to:

Select objects in many ways
ERASE unwanted entities
Restore items deleted in error
Draw circles, arcs and ellipses
Insert multiple lines of text
Insert centred lines of text
Define text styles using different fonts
Draw rectangle
Draw polylines with constant and varying widths
Merge polylines with arcs
Get Aerial Views
Create circles and doughnuts with solid fill
Select a hatch pattern and perform a test shading
Shade in multiple objects
Speed up REDRAWs

Chapter 5 CONSTRUCTIVE EDITING

General

The term constructive editing is used to describe those commands which either
replicate existing entities or alter their characteristics. This chapter begins with
some of the commands that alter the objects and then considers some of the
ways to duplicate them. This will help to accelerate the repetitive type of work
needed to complete the Eiffel tower (Figure 5.1) that was started in Chapter
3.

Start up AutoCAD, pick **File/** and find EXPRESS2.DWG from the exercise in Chapter 3. If you cannot find EXPRESS2.DWG, look for the backup
copy you made called TOWER.DWG. Once the drawing file has been loaded
in AutoCAD, pick **File/Save As...** and give the new name as **EIFFEL**.
AutoCAD automatically adds the "dwg" extension. This copies all of the EX-
PRESS2.DWG file to EIFFEL.DWG and leaves the old file unchanged.

If you haven't got the EXPRESS2.DWG file or another file containing
the drawing from Chapter 3, Table 5.1 summarizes what you will need. This
produces something similar to Figure 3.25 on page 67 and Figure 5.3.

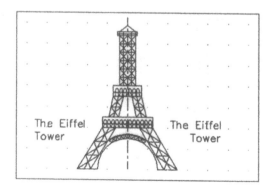

Figure 5.1 Eiffel target drawing

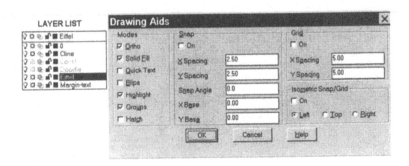

Figure 5.2 Drawing Environment

Table 5.1 EXPRESS2.DWG Summary

Layer	Status	Color	Linetype
0	ON	WHITE	CONTINUOUS
CLINE	ON	WHITE	CENTER
CONST	FROZEN	RED	CONTINUOUS
EIFFEL	ON	WHITE	CONTINUOUS
MARGIN-TEXT	ON	WHITE	CONTINUOUS

Current layer: EIFFEL
LIMITS from (0,0) to (65,45)   LTSCALE = 0.05

The center-line is drawn from (30,41.5) to (30,5) and is on the CLINE layer.
The following plines are all on layer EIFFEL.

Draw plines from (22.5,5) to (17.5,5) to (23.75,15) to (30,15)
From (23.75,15) to (23.75,17.5) to (30,17.5)
From (23.75,16) to (30,16)
From (25.3,17.5) to (26.5,22.5) to (30,22.5)
From (26.5,22.5) to (26.5,25) to (30,25)
From (26.5,23.5) to (30,23.5)
From (30,37) to (27.5,37) to (27.5,38) to (30,38)
From (27.5,37.5) to (30,37.5)
From (28,25) to (28,38) to (30,39.5) to (30,41.5)
From (29,37) to (29,38)

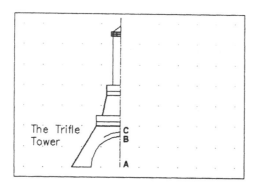

Figure 5.3 The Trifle Tower

Drawing the arch

Before engaging in all this editing let's do a bit more preparatory work on the tower. Set SNAP to 2.5, GRID to 5 by picking **Tools/Drawing Aids...** (Figure 5.2). Making sure that EIFFEL is the current layer, freeze the layers CONST and DOODLE if they are not already frozen. Use the pull-down layer list for this (Figure 5.2). Then, start by drawing the arch at the base of the tower. If you type the command you will have to input all the responses in bold but if you pick **Draw/Arc** followed by **Center, Start, Angle** from the pull-down menu you will only have to type the numbers.

Command: **ARC**
Center/<Start point>: **C** Center: **30,5**   (A)
Start point: **@7.5<90**   (B)
Angle/Length of chord/<End point>: **A** Included angle: DRAG **90**

There are lots of similar options on the ARC menu. Be sure you pick the correct one. Also, note that while AutoCAD does remember the last command it doesn't necessarily remember all the options within the last command. So, to draw the second arc, pick **Draw/Arc/Center, Start, Angle** again.

Command: **ARC**
Center/<Start point>: **C** Center: **30,5**   (A)
Start point: **@8.5<90**   (C)
Angle/Length of chord/<End point>: **A** Included angle: DRAG **30**

Add some text but this time use the Multi-line, MTEXT, command. Pick **Draw/Text** and **Multiline Text...** from the pull-down menu or pick the capital "A" icon from the tool-box.

Command: **MTEXT**
Current text style: STANDARD. Text height: 2.5
Specify first corner: **7,10**
Specify opposite corner or [Height/Justify...]: **20,15**

The Multiline Text Editor dialog box should appear (Figure 5.4). Select the
SIMPLEX font and a height of 1.8. Then type the text **The Trifle Tower**.
MTEXT automatically does the word wrap at the end ofthe line.

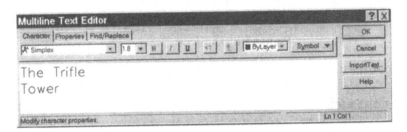

Figure 5.4 Multiline text

Finally to end the command pick the **OK** button. Don't worry that this
text has not been put on the MARGIN-TEXT layer. This sloppy practice will
be corrected later using AutoCAD's advanced editing facilities.

A small corner of the drawing will be used as the prep area where the
metalwork for the tower will be assembled. **ZOOM** in and draw the structure
panels. The zoom command can be abbreviated to **Z**. Type Z followed by
<ENTER>. AutoCAD will echo with the full name and proceed as normal.

Command: **Z <ENTER>** ZOOM
All/Center/.../Window/<Scale(X/XP)>: **W**
First corner: **2.5,2.5**
Other corner: **9,6.5**
Command: **SNAP**
Snap spacing or ON/OFF/Aspect/Rotate/Style <2.50>: **0.25**

The basic structure panel in Figure 5.5 is a cross-braced frame. It is first drawn
to fit within a 1 by 1 box so that it can later be scaled to fit the different parts of
the tower. While the coordinates of each point have been given in the command
sequence below, you may find it easier to simply pick the points with reference
to Figure 5.5. The current snap value facilitates this.

Command: **LINE**
From point: **3,5**   (A)
To point: **@1,1**   (B)

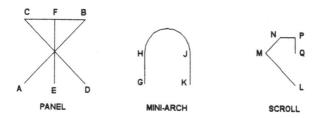

Figure 5.5 Structure panels

To point: **@−1,0**   (C)
To point: **@1,−1**   (D)
To point: **<ENTER>**
Command: **<ENTER>**
Line From point: **3.5,5**   (E)
To point: **@0,1**   (F)
To point: **<ENTER>**

Each landing will have a series of mini-arches comprising an arc and two lines
each. This is done using PLINE's line and arc modes.

Command: **PLINE**
Line From point: **5,5**   (G)
Current line-width is 0.00
Arc/Close/.../Width/<Endpoint of line>:**@0,0.5**   (H)
Arc/Close/.../Width/<Endpoint of line>: **ARC**   (switch to arc mode)
Angle/Center/.../Line/...<Endpoint of arc>: **@0.75<0**   (J)
Angle/Center/.../Line/...<Endpoint of arc>: **L**
Arc/Close/.../Width/<Endpoint of line>: **Length**
Length of line: **0.5**   (K)
Arc/Close/.../Width/<Endpoint of line>: **<ENTER>**

The final component to be used is the fancy iron scrolling for the large arch
at the base. For this you should use another PLINE.

Command: **PLINE**
From point: **7.5,5**   (L)
Current line-width is 0.0000
Arc/Close/.../Width/<Endpoint of line>: **@−0.5,0.5**   (M)
Arc/Close/.../Width/<Endpoint of line>: **@.25,.25**   (N)
Arc/Close/.../Width/<Endpoint of line>: **@.25,0**   (P)

Arc/Close/.../Width/<Endpoint of line>: @0,−.25   (Q)
Arc/Close/.../Width/<Endpoint of line>: <ENTER>

This should give you the three objects shown in Figure 5.5. As usual, the letters
on the diagram are only for reference and will not appear in your drawing.

Editing a polyline with PEDIT

Polylines are probably the most flexible entities in AutoCAD. This means
that they can also be cumbersome to edit. There are so many possibilities for
making changes that it is difficult to describe in a concise manner. In this
section you will encounter the more important facilities of Polyline EDITing
(PEDIT).

The PLINE scroll drawn above requires a few small changes. Firstly, it is
too small and secondly, it looks too square to be art nouveau. To make it a
bit bigger an extra point must be inserted and one vertex must be moved in
the polyline. Pick **Modify/Object** and **Polyline** from the menu bar. Pedit
can also be found on the Modify II toolbar.

Command: _pedit
Select polyline: **7.5,5**   (L)
Close/Join/Width/Edit vertex/.../eXit <X>: **Edit**

Choose the edit vertex option. The prompt line changes and an X appears on
the polyline at its first point or vertex.

Next/Previous/.../Width/eXit <N>: **N**   (M)

The **N** selects the next vertex on the polyline and the "X" should now be at
point M (7.0,5.5). Now type **Insert** to put in a new point.

Next/.../Insert/.../Width/eXit <N>: **I**
Enter location of new vertex: **@.25<90**   (R)

The new shape should look like Figure 5.6(b) with the X still at point M. Move
the X to vertex, N (7.25,5.75), by typing **Next** twice. Once the X is at the
correct vertex the vertex can be moved to its new location (Figure 5.6(c)).

Next/Previous/.../Width/eXit <N>: **N**   (N)
Next/.../Move/.../Width/eXit <N>: **M**
Enter new location: **@.25<90**   (S)
Next/Previous/.../Width/eXit <N>: **X**
Close/Join/.../Spline curve/.../Undo/eXit <X>: **S**
Close/Join/.../Spline curve/.../Undo/eXit <X>: **X**

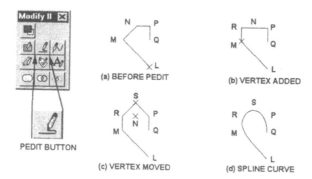

Figure 5.6 Polyline edit

Selecting the X option exits from the "Edit vertex" routine and returns you
to the PEDIT prompt, "Close/Join/...". To make the polyline into a smooth
curve select the Spline curve option by typing S. This executes a cubic B-spline
curve fitting routine. This type of curve gives an extremely smooth shape fitted
to the vertex points of the orginal polyline. The curve won't actually pass
through the vertices but will be drawn nearby. The technique is named after
the mathematician, Bezier, who invented it. Bezier is regarded by many as the
father of computer graphics and surface modelling.

Early versions of AutoCAD gave only the "Fit" option for smoothing
polylines. This just changes all the line segments into arc segments. The arcs
all pass through the original vertices. It is faster but visually inferior to the
spline.

Other options for editing polylines include making it into a closed polyline,
joining two polylines together and changing the width. You can also "Decurve"
a spline or fitted curve.

The scroll work should now resemble the curlicue shown in Figure 5.6(d).

An alternative to creating a polyline and then editing it is to use the
SPLINE command, introduced in AutoCAD Release 14. Here is a quick run
through, with a comparision with the polyline.

Command: **SPLINE**
Object/<Enter first point>: **7.5,5**   (L)
Enter point: **@−0.5,0.5**   (M)
Close/Fit Tolerance <Enter point>: **@0,0.25**   (R)
Close/Fit Tolerance <Enter point>: **@0.25,0.25**   (S)
Close/Fit Tolerance <Enter point>: **@0.25,−0.25**   (P)
Close/Fit Tolerance <Enter point>: **<ENTER>**
Enter start tangent: **<ENTER>**

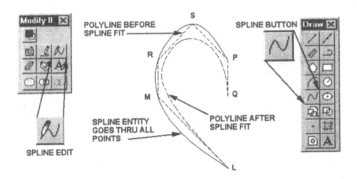

Figure 5.7 Polyline compared with Spline

Enter end tangent: <**ENTER**>

Note that the spline in Figure 5.7 passes through all the points while the curved polyline is only close to them. Your screen should show only the two curves. The original polyline is included for comparison. Also, we missed out the point, Q, when drawing the SPLINE. There are a few other subtle differences in the way the curves are calculated but this gives you the important points. The SPLINE command is an excellent tool for curve fitting for graphs and contours.

Moving objects

All AutoCAD entities can be modified to alter their position in the drawing. This gives great flexibility when making a drawing as you don't have to worry about getting everything to fit exactly. As the drawing progresses conflicts can be resolved by moving the objects to give a clearer picture.

You can zoom out using the Aerial view button (airplane icon) for a better view and move the panels into position. See Figure 4.24 (page 102) if you need help in doing this. Alternatively try, ZOOM All followed by a suitable window.

The X-Y axes icon (UCS icon) in the lower left corner of the drawing may now obscure the view of small panels. To remove the icon pick **View/Display/UCS Icon**. This shows a short menu with a tick in front of the "On" item. Pick **On** and both the tick and the UCS icon will disappear. You can also do this by typing the command UCSICON.

Pick **MOVE** icon from the Modify toolbar (Figure 5.8). The icon looks like a double headed arrow pointing right and left. You will also find this command in the Modify pull-down menu. AutoCAD enters the "Select objects" mode and you can make a window around the mini-arch. When all the selections have been made you will be asked how far you want to move it.

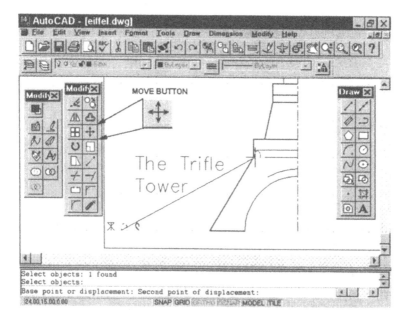

Figure 5.8 Moving the mini-arch

Command: **MOVE**

  Select objects: **5,5**

  1 found

  Select objects: **<ENTER>**

  Base point or displacement: **5,5**

This is the point from which the object is to be moved. You are then prompted for the new location. The mini-arch is then moved to the first landing (Figure 5.7).

  Second point of displacement: **24,15**

The base point and second point do not have to be at the old and new locations. All that is important is the relative displacement between the two points. For example, the above movement would also have been achieved if the points (0,0) and (19,10) were input.

  Base point or displacement: 0,0

  Second point of displacement: 19,10

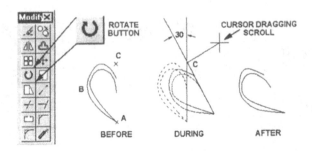

Figure 5.9 Rotating the scroll

A third way of achieving the same result is to input the relative displacement at the first prompt and just press **<ENTER>** at the second.

>   Base point or displacement: 19,10
>   Second point of displacement: <ENTER>

Of course you can also use the mouse to drag the object to the correct location. This can involve the use of object snapping.

Rotating objects

Before moving the scroll work to the main arch it must be put into the correct orientation. The scroll will be used to fill the area between the shorter and longer arcs. As the shorter one has an included angle of 30 degrees the scroll has to be rotated by that angle (Figure 5.9). To see the operation clearly, zoom in closely. As it is not so easy to pick small windows with the aerial viewer we will use the standard ZOOM command.

>   Command: **ZOOM**
>   All/Center/.../Window/<Scale(X/XP)>: **W**
>   First corner: **6,4**
>   Other corner: **9,7**

The ROTATE command is also on the Modify pull-down menu and also on the tool box. Click the **Rotate** button from the tool box. The object selection procedure is as before and when all the objects have been found you are asked for the center of rotation and the angle.

>   Command: **ROTATE**
>   Select objects: **7.5,5**   (Pick the polyline at A.)
>   1 found

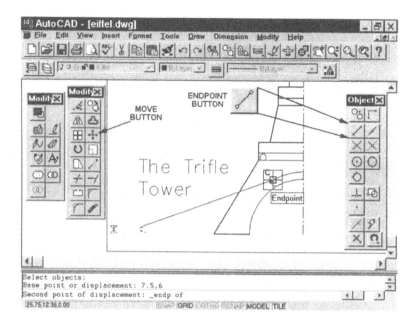

Figure 5.10 The scroll in the arch

Select objects: **7,5.5**   (Pick the spline at B.)
1 found
Select objects: **<ENTER>**
Base point: **7.5,6**   (Center of rotation at C.)
<Rotation angle>/Reference: **30**

The scroll has been rotated 30 degrees anti-clockwise about the point (7.5,6).
Now do a "ZOOM Previous" and move the polyline to the end of the shorter
arc. The point (7.5,6) will be used as an imaginary handle by which the curve
will be moved to the point C shown in Figure 5.10. Use object snap endpoint
to pick the end of the arc as shown below.

Command: **Z ZOOM**
All/.../Previous/.../<Scale(X/XP)>: **P**
Command: **MOVE**   (Pick the MOVE button.)
Select objects: **W**
First corner: **6,5**
Other corner: **9,7**
2 found.
Select objects: **<ENTER>**
Base point or displacement: **7.5,6**

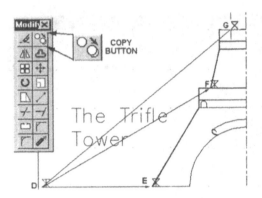

Figure 5.11 Multiple copies

Second point of displacement: **end** of Pick upper arc at point C.

Copying

The basic structure panel is still on the ground awaiting erection. Instead of simply moving it over you can copy it to each tier of the structure (Figure 5.11). The COPY command operates a bit like MOVE. You first select the objects to copy and then provide a displacement showing where you want them copied to. Unlike MOVE the original objects are left untouched and duplicates appear in the new positions.

In this particular operation you will make multiple copies of the panel, an extra feature of COPY. The first copy will be made to point E at the base, the second to point F and the third to the upper tier at G. The base point for all three copies will be the lower left corner of the panel, D (3,5).

Command: **COPY**
Select objects : **Window**
First corner: **3,5**   (D)
Other corner: **@1,1**
4 found

This surrounds the panel and finds its four lines.

Select objects: **<ENTER>**
<Base point or displacement>/Multiple: **Multiple**
Base point: **3,5**   (D)

Second point of displacement: **17.5,5**   (E)

The panel should now appear at the point E and the prompt asks for "Second point..." for the next copy. This uses the same base point. Pick the **intersection** button from the tool box (looks like an X on the second last row) before picking the point F.

Second point of displacement: **int** of Pick point F.

The third copy is made to point G (28,25)

Second point of displacement: **28,25**   (G)
Second point of displacement: **<ENTER>**

When all the desired copies have been made press **<ENTER>** without giving any "Second point...".

The COPY command can also be used to make single copies. To do this give the coordinates of the "Base point or displacement" instead of typing "Multiple". It will then work as above but will ask you for a single "Second point..." before returning to the "Command:" prompt.

To draw the "inside leg" of the tower (Figure 5.12) you can copy the inclined lines. Copy the line EH to a point 3.3333 units to the right and the leg of the middle tier 2.5 to the right.

Command: **COPY**
Select objects: Pick line EH. 1 found.
Select objects: **<ENTER>**
<Base point or displacement>/Multiple: **3.3333,0**
Second point of displacement: **<ENTER>**
Command: **<ENTER>**
COPY
Select objects: Pick the long inclined line passing through F. 1 found.
Select objects: **<ENTER>**
<Base point or displacement>/Multiple: **2.5,0**
Second point of displacement: **<ENTER>**

Notice anything wrong? In Chapter 3 we created the leg EH as part of a polyline QEHJ. Here we have copied the whole polyline. To get rid of the bits we no longer need H'J' and E'Q' use the **Explode** command and Erase them. Pick the new polyline Q'E'H'J' and then pick the Explode button. This breaks the polyline into three LINE entities.

Command: **Explode**
Select objects: **32,15**
1 found.

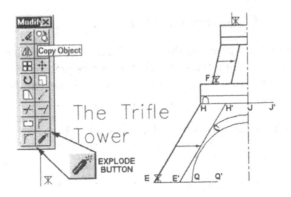

Figure 5.12 The inside leg

Select objects: **<ENTER>**
Command: **ERASE**
Select objects: **32,15**
1 found.
Select objects: **24,5**
1 found.
Select objects: **<ENTER>**

Often in CAD the easiest way of drawing items is to copy more than you need
and then delete the unwanted details.

Altering objects' characteristics

All the basic pieces of the tower are now in place. Something must be done to
correct their sizes to make them all fit. The panels are too small and there is a
gap between the short arc at C and the inner part of the leg. Also the original
panel is still at the point (3,5) on the Eiffel layer.

The first bit of tidying up is to sort out the use of layers. The text was put
on the EIFFEL layer but ideally it should have been on MARGIN-TEXT. Also
there was a small spelling error in the name of the tower! Can one command
rescue the situation?

Pick **Modify/Properties…** from the pull down menu. This runs a context sensitive command with the wierd name of "_ai_propchk". This responds
with different dialog boxes depending on how many objects and the type of
object selected. It is another name for AutoCAD's **DDMODIFY** command. To
correct the spelling error pick the "The Trifle Tower" text in the drawing.

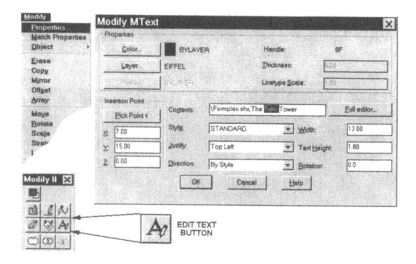

Figure 5.13 Modify Text dialog box

Command: _ai_propchk
Select objects: Pick the letter "f" in the word Trifle
Select objects: **<ENTER>**

This command then brings up a context-sensitive dialog box. As the entity
selected is a multiline text entity the Modify MText dialog box pops up (Figure 5.13). This contains all the parameters defining the text. Highlight the
word **Trifle** and replace it with bf Eiffel. Then pick the **Layer** button. From
the resulting "Select Layer" dialog box, pick **MARGIN-TEXT** followed by
OK. Pick **OK** from the Modify MText dialog to execute the changes. The
old text is then replaced by the new (Figure 5.14). Note that the button "Full
editor" would bring up the standard Multiline text editor.

Above we needed to change two of the text entity's parameters, namely its
layer and its value. We could also have altered the size, justification, position,
etc. If you just want to correct a spelling error there is another command
called DDEDIT. There is also a Text Edit button on the Modify II toolbar.
This brings up the standard Multiline text editor.

The easiest way to change the layer of an object is to select it and then
pick the target layer from pull down list on the Object Properties toolbar.
To move the original panel from the Eiffel layer to the CONST layer, select
the four line entities either by picking a window to surround them or picking
them individually. You can use the implied windowing feature by picking the
corners when the Command prompt is shown. Then go to the layer list and
pick **CONST** as shown in Figure 5.14. Once this is done you will get a prompt

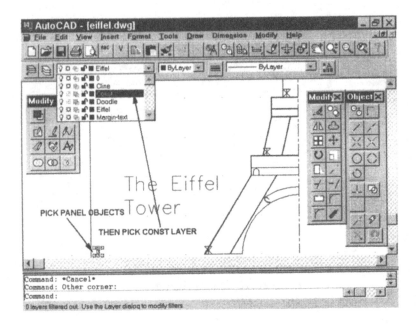

Figure 5.14 Changing text and layers

that the items are being transferred to a frozen layer and will be removed from the selection set. Then pick **OK** to set the new layer for the four entities. Their color and linetype will automatically be updated to the layer settings.

Note for AutoCAD 12 users Two similar commands are still available; DD-CHPROP and the non-dialog box version of this command called, CHPROP, which users of older versions of AutoCAD will be familiar with. For example, to do the previous operation you would type:

Command: **CHPROP**
Select objects: **Window**
First corner: **3,5**
Other corner: **@1,1** 4 found
Select objects: **<ENTER>**
Change what property Color/LAyer/LType/Thickness) **LAYER**
New layer <EIFFEL>: **CONST**
Change what property (Color/LAyer/LType/Thickness) **<ENTER>**

In EIFFEL there is a gap between the inner part of the tower's leg and the shorter of the two arcs in the arch. The EXTEND command can be used to change the arc's end point so that it meets the line exactly (Figure 5.15).

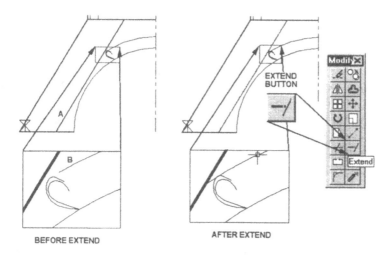

Figure 5.15 Extending the arch

Having picked **EXTEND** from the "Modify" menu you then must give Auto-CAD the boundary edges to which you want to extend the entity. In this case pick the inner line.

Command: **EXTEND**
Select boundary edges: (Projmode = UCS, Edgemode = No extend)
Select objects: **22,7**   or pick the line near (A)
1 found

You will then be prompted for a second edge. In this example, only one edge is required. AutoCAD then asks for the objects to be extended. Pick the shorter arc and the curve is projected to meet the line.

Select objects: <**ENTER**>
<Select object to extend>/Project/Edge/Undo: **26.5,12.75**   (B)
<Select object to extend>/Project/Edge/Undo: <**ENTER**>

You can extend more than one item so AutoCAD prompts for more. Pressing <ENTER> without making another selection exits back to the "Command:" prompt.

While EXTEND is used to project to a new intersection point the TRIM command is used for objects that are already crossing. TRIM works similarly to EXTEND; the boundary edges are selected first and then the objects to trim.

Enlarging objects

In order to fit, each of the tower leg structure panels must be enlarged. The top panel, near the point G must be increased by a factor of 2, the middle panel near F by a factor of 2.5 and the bottom one by 3.3333. The SCALE command allows you to increase the dimensions of an object (Figure 5.16). Its operation is similar to that of ROTATE. You select the objects to SCALE and give a base point about which the objects will move when enlarging. Finally, you specify the magnification factor. The factor must be positive and not equal to zero. Giving a factor less than 1 reduces the size of the object while values greater than 1 increase it. The X and Y dimensions are changed by the same amount.

Using an initial panel size that fitted within a 1 by 1 box makes the calculation of the appropriate scale factors straightforward. To enlarge the panel at point G pick **SCALE** from the Modify menu or use the button on the Modify toolbar. Then select the panel using **window**, give point G as the base point and finish up by giving the scale factor of **2**.

Command: **SCALE**
Select objects: **Window**
First corner: **28,25**   (G)
Other corner: **@1,1**
4 found
Select objects: <**ENTER**>
Base point: **28,25**
<Scale factor>/Reference: **2**

Now scale the panel at F. Pressing <ENTER> re-executes the last command. Use object snap intersection to locate the base point at F.

Command: <**ENTER**>
SCALE
Select objects: **Window**
First corner: **25,17**   (near F)
Other corner: **27,19**
4 found
Select objects: <**ENTER**>
Base point: **INTERSEC** of Pick point F.
<Scale factor>/Reference: **2.5**

And finally to scale the bottom panel. This time use the intersection points at E and J to specify the scale factor.

Command: <**ENTER**>

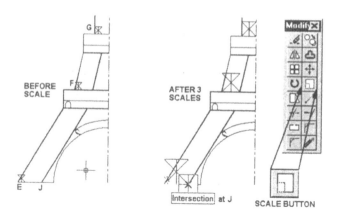

Figure 5.16 Scaling up the panels

SCALE
Select objects: **Window**
First corner: **17.5,5**   (E)
Other corner: **@1,1**
4 found.
Select objects: **<ENTER>**
Base point: **17.5,5**   (E)
<Scale factor>/Reference: **R**
Reference length <1>: **<ENTER>**
New length: **INT** of Pick point J.

The reference length is the length of the original object. It can be picked by giving two points on the object. In this case the default length of 1 is correct.

Stretching objects into shape

The two lower panels are now the required size but they are the wrong shape for the inclined legs. To fit they must be changed to become skew with the angle of the respective leg. Concentrating on the bottom panel first, zoom in for a closer view. To make it skew, the top of the panel must be moved sideways while the bottom stays put. That is, all the line end points in the top half will be shifted to the right by the distance from S1 to S2 (Figure 5.17). The STRETCH command, like EXTEND has the ability to act on one end of an entity while leaving the other end untouched.

STRETCH is easiest to operate if it is picked from the Modify menu. When you pick **STRETCH** you will be given the message that you must select the objects using a window and the "Crossing" option is automatically chosen. The "Crossing" option picks up all objects totally or partially within the window. This window has to contain all the end points to be shifted. You can use implied crossing window by picking the top right corner first or you can be explicit and type **C** for crossing at the Select objects prompt.

Command: **STRETCH**
Select objects to stretch by crossing window or crossing polygon ...
Select objects: **c**   (Crossing)
First corner: **21.5,9**   (W1)
Other corner: **17,7.5**   (W2)
5 found

Even the inclined leg becomes ghosted because it crossed through the window. As only the end points that were actually within the window are stretched, the leg will be okay. However, if you are in doubt as to whether an object might be undesirably stretched you can **Remove** it from the selection set. For practice remove the leg.

Select objects: **Remove**
Remove objects: **21.5,11.5**   (Point R1 on leg)
1 found, 1 removed
Remove objects: **<ENTER>**

When the selection/de-selection has been completed you are prompted for the point to stretch from and for the new destination.

Base point or displacement: **INT** of Pick point S1.
Second point of displacement: **INT** of Pick point S2.

The panel should now fit snugly into the inclined leg (Figure 5.17).
   Repeat this procedure to make the panel at point F fit its leg. Remember that only the end points that are within the window are stretched.

Command: **STRETCH**
Select objects to stretch by window or polygon ...
Select objects: **28,20.5**   (W3 Implied Crossing)
Other corner: **25,19.5**   (W4)
5 found.
Select objects: **<ENTER>**
Base point or displacement: **INT** of **25,20**   (near the top left corner)
Second point of displacement: **INT** of **26,20**   (Intersection with leg)

The COPY command has already been used to make multiple copies of the original structure panel. It will now be used to duplicate the panel on the

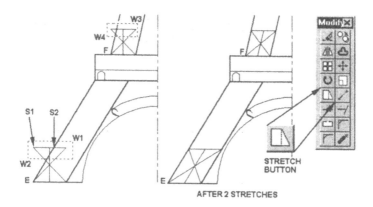

Figure 5.17 Stretching some points

middle tier near point F. The other tiers require multiple copies of their panels. As these copies are in regular patterns, AutoCAD's **ARRAY** command will be used. This allows objects to be copied in rows and columns and circular arrays. The scroll work will first be mirrored to complete the heart shape and copied in a circular array along the arch.

To copy the panel at F use **COPY Window** to select and use object snap **intersec** at points **F**, near (25.25,17.5), and near (26,20), to give the displacement.

Command: **COPY**
Select objects: **25,17**   (Using implied windowing again)
Other corner: **28.5,20.5**
4 found
Select objects: <ENTER>
<Base point or displacement>/Multiple: **INT** of Pick point F.
Second point of displacement: **INT** of Pick point (26,20).

Mirror image

Mirroring objects allows the user to take advantage of symmetries in the object being drawn. For example, only half of the tower is being drawn since it is symmetrical about the center-line. At the end of the chapter it will be mirrored to complete the picture.

To illustrate the MIRROR command the scroll in the arch will be reflected in a line at an angle of 120 degrees (remember that is was previously rotated by 30 degrees from the vertical and 30+90=120). Before mirroring zoom in for a better view. Pick **Mirror** from the **Modify** menu or toolbar and select the scroll polyline and spline. For the mirror line give the intersection point of the polyline and the lower arc as the first point and a relative displacement with an angle of 120 degrees for the second.

> Command: **ZOOM**
> All/Center/.../Window/<Scale(X/XP)>: **W**
> First corner: **22.5,10**
> Other corner: **29,14.5**
> Command: **REGEN**   (To refresh the screen)
> Command: **MIRROR**
> Select objects: **25,11**   (Window around the two curves)
> Other corner: **27,13**
> 2 found
> Select objects: **<ENTER>**
> First point on mirror line: **INT** of Pick point M1, Figure 5.18
> Second point: **@10<120**

The length of the mirror line doesn't matter, just its orientation. You can also drag the mirror line until its orientation is correct.

> Delete old objects? **<N> <ENTER>**

You do not wish to delete the original, so press **<ENTER>** and the heart is complete (Figure 5.18). At the end do a **ZOOM Previous** to get back to the last display magnification.

> Command: **Z**
> ZOOM
> All/.../Previous/.../<Scale(X/XP)>: **P**

Multiple copies using ARRAY

In the tower drawing there are two simple patterns to be copied. Firstly, the mini-arches at the first landing will be copied to give two rows and six columns. Then the panels in the top section will be copied to give six rows and one column. Two slightly more difficult operations are involved to copy the hearts along the arch and the panels along the sloping lower leg.

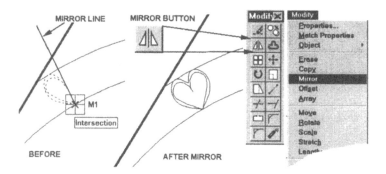

Figure 5.18 Taking polylines to heart

Rectangular arrays

With the ARRAY command you select the original objects to be arrayed. Then you specify whether they are to be copied in a rectangular grid or circular pattern and finally you give the dimensions of the pattern's repeated unit.

To make the rows of mini-arches in Figure 5.19, pick **Array** from the Modify menu or toolbar and select the polyline mini arch.

Command: **ARRAY**
Select objects: **24.75,15.5**   (Point A on right hand leg)
1 found
Select objects: **<ENTER>**
Rectangular or Polar array (<R>/P): **R**

This selects the rectangular grid pattern. You are now prompted for the number of rows and columns.

Number of rows (---) <1>: **2**
Number of columns (| | |) <1>: **6**

The "(---)" is to remind you that the rows are always horizontal and the "(| | |)" is for the vertical columns. Now input the distance between the rows and the columns.

Unit cell or distance between rows (---): **1**
Distance between columns (| | |): **1**

The mini-arch should be repeated to give a total of 12 arches. Note that the distance between the rows and columns is the length between a point on the original object and the corresponding point on its immediate neighbours.

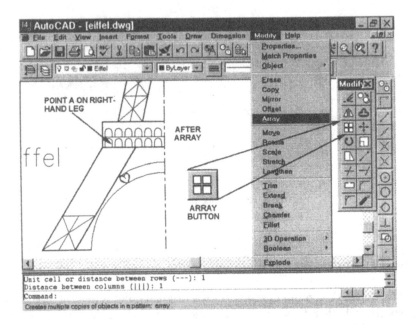

Figure 5.19 Rectangular array

Inputting positive distances causes the duplicates to appear to the right and above the original. To make them appear on the left give a negative distance between rows. Similarly a negative distance between the columns causes the new objects to be drawn below the original.

To ARRAY the panel in the top part of the structure try the following:

Command: **Zoom**
All/.../Previous/.../<Scale(X/XP)>: **ALL**
Command: **ARRAY**
Select objects: **Window**
First corner: **28,25**   (G on Figure 5.20)
Other corner: **@2,2**
4 found
Select objects: **<ENTER>**
Rectangular or Polar array (R/P)<R>: **<ENTER>**
Number of rows (---) <1>: **6**
Number of columns (|||) <1>: **<ENTER>**
Unit cell or distance between rows: **2**

This should fill up the top of the structure with a total of 5 copies plus the original panel.

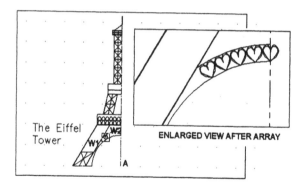

Figure 5.20 Polar array

Circular arrays

The circular or polar array option is used to copy objects around some central focus point. For example, the spokes on a bicycle wheel could be drawn by copying one line in a circular pattern centerd on one of the end points. In the EIFFEL tower drawing the heart shapes will be copied along the arch. This can be done by using a polar array centerd on the point O (at the arc center) and repeating the object through 30 degrees, Figure 5.20.

Zoom in for more detail. Pick **ARRAY**, select the two halves of the heart shaped scroll and then opt for the **Polar** array.

```
Command: Zoom
All/Center/.../Window/<Scale(X/XP)>: W
Firsr corner: 17,4
Other corner: 32,15
Command: ARRAY
Select objects: W
First corner: 25,11                                          (W1)
Other corner: 27,13                                          (W2)
4 found
Select objects: <ENTER>
Rectangular or Polar array (R/P): P
```

You are now prompted for the center point of the array, the number of items to be in the array (copies plus the original) and the number of degrees to fill.

```
Center point of array: 30,5                                  (A)
Number of items: 5
Angle to fill (+=ccw, −=cw) <360>: −30
```

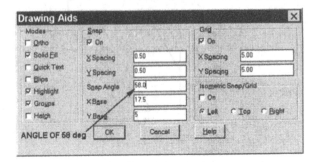

Figure 5.21 Snap angle

Rotate objects as they are copied? <Y> <**ENTER**>

The minus sign indicates a clockwise angle (as indicated by the "−=cw" in the parentheses). A positive angle would cause the copies to appear anti-clockwise from the orginal. The final prompt asks you if the objects are to be copied in their current orientation or if they are to be rotated. In this case they should be rotated. If you were drawing something like the carriages in a ferris wheel then they would not be rotated.

Rotated rectangular arrays

In rectangular arrays above, the rows were vertically above each other. Similarly the columns are separated by horizontal distances. This is because the ARRAY directions are always parallel and perpendicular to the SNAP angle. The default snap angle is zero. Using **Tool/Drawing aids...** you can rotate the snap angle to 58 degrees (the angle of the lower leg) and change the snap base to 17.5,5 (the bottom of the leg) as shown in Figure 5.21. The reason for changing the Snap base point is that when the grid is rotated by 58 degrees, the bottom of the leg no longer lies on a snap point.

To use the dialog box you must already know the values for the snap angle and base point. If you need to align these with objects in the drawing you will have to type the SNAP command and then use the ROTATE option.

Note that cursor cross-hairs and grid rotate by 58 degrees (Figure 5.22). Note also that the drawing coordinates have not changed. It's just the snap locations that have altered. You can use ARRAY to copy along the 58 degree leg. This array has one row and three columns as the leg direction is the new "horizontal" direction. The distance between the columns can be found by object snapping to points E and N.

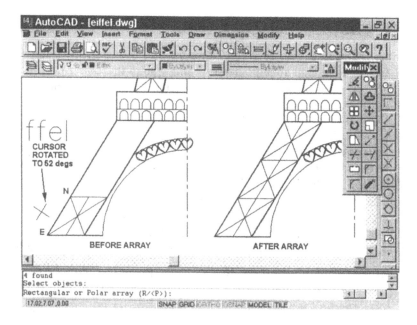

Figure 5.22 Rotated rectangular array

Command: **ARRAY**
Select objects: **Window**
First corner: **17.5,5**   (E)
Other corner: **23,9**   (Note the selection window is not rotated)
4 found.
Select objects: **<ENTER>**
Rectangular or Polar array (R/P)<P>: **R**
Number of rows (---) <1>: **1**
Number of columns (| | |) <1>: **3**
Distance between columns: **INT** of Pick point E.
Second point: **INT** of Pick point N.

Now reset the SNAP angle back to zero.

Command: **SNAP**
Snap spacing or ON/OFF/Aspect/Rotate/Style <0.50>: **Rotate**
Base point <17.50,5.00> **0,0**
Rotation angle <58.00>: **0**
Command: **Zoom**
All/Center/.../Window/<Scale(X/XP)>: **A**

Finishing up

The tower is now near completion. All that remains is to copy the mini-arches to the second level and to mirror everything about the center-line. As this particular operation involves a lot of entities it will be fairly computationally intensive.

> Command: **COPY**
> Select objects: **Window**
> First corner: **24,15**   (Selecting 6 arches)
> Other corner: **26.75,17**
> 6 found
> Select objects: **<ENTER>**
> <Base point or displacement>/Multiple: **24,15**
> Second point of displacement: **27,22.5**   (In middle landing)

Before executing any big copying operation you should save the drawing. Pick **File/Save...** from the pull down menu or pick the Save button on the Standard toolbar. This executes a "quick save". This saves to the current file name without requesting a confirmation.

> Command: _qsave

Now you can mirror the left half of the tower including the text.

> Command: **MIRROR**
> Select objects: **Window**
> First corner: **5,5**
> Other corner: **31,42**
> 99 found

If the actual number of objects found on your drawing is much greater than 99 it means that you probably have copied some items onto themselves. Not to worry as we are nearly at the end of the exercise anyway. However, in order to avoid unnecessary objects being mirrored you can remove the center line and the rightmost heart from the selection set.

> Select objects: **R**
> Remove objects: **30,5**   (The center line)
> 1 found, 1 removed
> Remove objects: **29.25,12**
> Other corner: **31,14**
> 4 found, 4 removed
> Remove objects: **<ENTER>**
> First point on mirror line: **30,5**   (A)
> Second point: **@35<90**

Figure 5.23 The completed Eiffel tower

Delete old objects? <N> <**ENTER**>

Again, the length of the mirror line doesn't matter, just its location and direction.

Is your picture like Figure 5.23 or did the text become inverted? Auto-CAD supports two modes of mirroring text. The normal mode is for it to be inverted like everything else. Many times this can yield silly results. The second mode which prevents all text from becoming reversed is invoked by setting an AutoCAD system variable. The variable is called MIRRTEXT and when its value is set to zero the mirrored text will not be inverted.

Command: **MIRRTEXT**
New value for MIRRTEXT <1>: **0**

Now ERASE the inverted text and repeat the MIRROR command for the original text only.

Command: **ERASE**
Select objects: **45,15**   (Approx. top line of text)
1 found
Select objects: <**ENTER**>
Command: **MIRROR**
Select objects: **14,15**   (top line of text)
1 found
Select objects: <**ENTER**>
First point on mirror line: **30,5**

     Second point: **@1<90**
     Delete old objects? **<N> <ENTER>**

Note that the MIRROR command preserves the layer of the entity. So even though you were working with EIFFEL as the current layer the new text will be puty on the Margin-text layer. Furthermore the copied items are always put on the same layer as their original.

    To finish the session, save the drawing and to exit AutoCAD type **END** or pick **File/Exit** from the pull-down menu and pick **Yes** to the the **Save Changes** prompt.

    Command: **END**

Summary

In this chapter you have been introduced to most of AutoCAD's editing commands. Some new aspects of other commands have also been covered.

    You should now be able to:

Edit polylines and create smooth curves.
Move, rotate and copy objects within a layer.
Make multiple copies of objects in regular patterns.
Alter text.
Move things from one layer to another.
Change the proportions of objects.
Use the MIRROR command.
Set the MIRRTEXT variable.

Chapter 6 SUPER-ENTITIES

General

A number of entities can be grouped together to form a single new super-entity. In a way polylines are super-entities since they cause a number of lines and arcs to be grouped together. AutoCAD provides a more general way of linking objects together to form *blocks*. Blocks are mini-drawings that can be called up or inserted into a drawing at any location and as many times as desired. By compounding entities to form frequently used shapes or symbols the AutoCAD user can avoid unnecessary duplication of drafting. Blocks can be made globally available to drawings allowing AutoCAD users to build up libraries of complicated shapes which can be easily incorporated into any new drawing. Indeed, the true benefits of AutoCAD only become apparent when you have such libraries set up. Large assembly drawing can be quickly created by inserting the standard details from your library.

Blocks are more than just a stored shape. They can also contain non-drawing information such as a part number or a cost for the item. This extra information can be accessed to provide a bill of materials which could significantly improve the accuracy and speed of your estimates for the job.

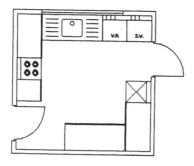

Figure 6.1 Kitchen fitted with blocks

Table 6.1 Layer settings for drawing KITCHEN

Layer name	State	Color	Linetype
0	On	7 (white)	CONTINUOUS
CONST	On	2 (yellow)	CONTINUOUS
FITTINGS	On	7 (white)	CONTINUOUS

Current layer: 0

Making a block

The exercise in this chapter uses the fitted kitchen industry to illustrate how to create and use the blocks in Figure 6.1. This will also cover adding text information to blocks and extracting this information. While making up some suitable objects for AutoCAD Express Kitchens Inc. you will encounter a couple of new editing commands, FILLET and OFFSET and a really neat way to draw parallel lines. The chapter finishes off with a look at some of AutoCAD's inquiry commands which give information about the drawing and also about the computer.

To start with you should begin a new drawing, using Metric default setting and calling it KITCHEN. Use the **Save** command to give the filename. The operating unit for the drawings in this chapter is the millimetre and so the limits must be set for an upper right corner of (6500,4500). The layers should be set up as given in Table 6.1. When this is done and the snap, grid and axis are set you can begin on the first block. This is to be the symbol for a door and comprises some lines and an arc.

Command: **LIMITS**
Reset Model space limits:
ON/OFF/<Lower left corner> <0.0000,0.0000>: <**ENTER**>
Upper right corner <420.0000,297.0000>: **6500,4500**
Command: <**ENTER**>
LIMITS
Reset Model space limits:
ON/OFF/<Lower left corner> <0.0000,0.0000>: **ON**
Command: **Z**
ZOOM
All/Center/.../Scale(X/XP)/Window/<Realtime>: **A**

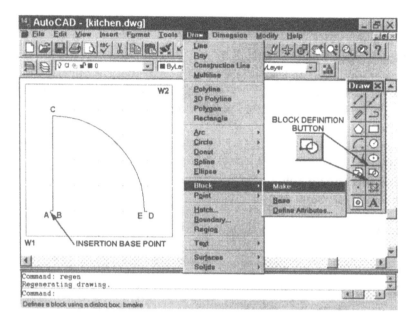

Figure 6.2 Blocking the door

Now set the snap value to 100 and the grid at 200 and change the units precision to 1 place after the decimal point.

Command: **SNAP**
Snap spacing or ON/OFF/Aspect/Rotate/Style <10.0000>: **100**
Command: **GRID**
Grid spacing(X) or ON/OFF/Snap/Aspect <0.0000>: **2X**
Command: **DDUNITS**   (Or pick Format/Units...)

The "2X" response for Grid spacing means twice the current snap setting. Use **DDUNITS** to bring up the Units Control dialog box and select **0.0** from the Precision pull down list. Refer to Figure 2.8 on page 22 if you have difficulty with this.

A standard size door in a dwelling house has an opening approximately 800mm wide. Allowing for the frame, the door itself will make an arc with a radius of 750mm (Figure 6.2).

Command: **LINE**
From point: **1000,1000**   (A)
To point: **@25,0**   (B)
To point: **@0,750**   (C)

To point: <ENTER>
Command: **ARC**
Center/<Start point>: **@**   (This selects the last point, C.)
Center/End/<Second point>: **C**   (The center point, B.)
Center: **1025,1000**   (The intersection of the two lines.)
Angle/Length of chord/<End point>: **Angle**
Included angle: **-90**
Command: **Zoom**   (For a closer look at the door)
All/Center/.../Scale(X/XP)/Window/<Realtime>: **W**
First corner: **500,500**
Other corner: **3500,3500**

The negative angle for the ARC is required since the positive direction for
angles is anti-clockwise.

Command: **LINE**
From point: **1800,1000**   (D)
To point: **@-25,0**   (E)
To point: <ENTER>

This is a door which is hinged on its left-hand side. The arc indicates the area
swept by the door as it is opened. The assembly is now ready for BLOCKing
(Figure 6.2). Pick **Draw/Block** from the menu bar followed by **Make**....
This executes a command called **BMAKE**, short for Block-make. The Block
Definition dialog box appears (Figure 6.3). Give the block name as **DOOR**
and pick the **Select point** to choose the base point as the point **1000,1000**
(point A on Figure 6.2). The insertion base point is the reference point on
the object by which it will be located. It is in effect the origin point for the
block. Then pick the **Select Objects** button to select the drawn entities to
be included in the block.

Command: **BMAKE**
Insertion base point: **1000,1000**   (A)
Select objects: **800,800**   (W1)
Other corner: **2000,2000**   (W2)
4 found
Select objects: <ENTER>

The four objects, three lines and an arc should become ghosted indicating that
they have been selected. The Block Definition dialog reappears when you end
the selection procedure. The number found should be shown as "4" on the
dialog (Figure 6.3). Remove the check in the "Retain objects" box and pick
OK. The 4 items will disappear from the screen and into the list of blocks.

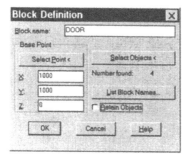

Figure 6.3 Block definition

The rules for naming blocks are the same as for naming LAYERS: up to 31 characters long containing letters, numbers and the characters "$", "-" and "_". No spaces are allowed in the block name. If you break these rules the dialog box will prompt "Invalid block name" just below the OK button when that button is picked. You can change the name and carry on.

Inserting blocks

To draw the block pick **Insert/Block**... from the menu bar. This is almost the reverse of the blocking procedure. A dialog box pops up (Figure 6.4) and you are asked for a block name to insert and where to place its insertion base point. There is considerable flexibility with inserting blocks. You can position it anywhere. You can alter its scale in either the X or Y direction and also change its orientation.

For the first insertion, pick the **Block**... button and pick the **DOOR** followed by **OK**. Now, uncheck the "Specify Parameters..." in theOptions part of the Insert dialog box and give the X and Y insertion point values as **2000** each. Then pick **OK**. This should give you the door marked "Insertion A" on Figure 6.5.

For "Insertion B" on Figure 6.5 we will use the same block but specify the points on the screen. This is probably the most common method of input as it gives you flexibility and access to object snap for locating the blocks. Pick **Insert/Block**... once more. Note that AutoCAD remembers the block name from the most recent insertion. Now, check the **Specify Paramters on Screen** box. The insertion point, scale and rotation fields become greyed out. Pick the **OK** button. The dialog box disappears and the new door is dragged around with the cursor. Input the values in the following sequence.

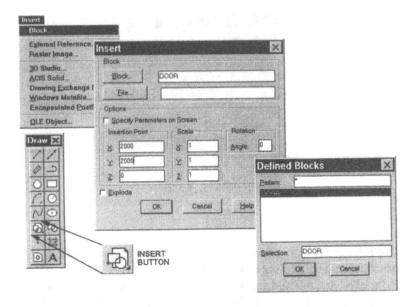

Figure 6.4 Insert menu and dialog box

Command: _ddinsert
Insertion point: **4000,2000**
X scale factor <1> / Corner /XYZ: **−1**   (This flips the door over)
Y scale factor (default =X: **1**   (To avoid flipping vertically)
Rotation angle <0>: **<ENTER>**

Note that the change to the X scale is automatically applied to the Y unless
we override it. A scale greater than 1 increases the size of the block, a scale
between 0 and 1 reduces the size while negative values flip the block over. The
rotation can be specified by moving the cursor and picking or by typing the
angle.

Finally, to insert the door at C in Figure 6.5 at right angles to the other
two use a rotation angle of 90 degrees (positive = anti- clockwise). This time,
rather than use the dialog box try inputting the information with the keyboard.
At the command prompt type **INSERT** as follows:

Command: **INSERT**
Block name (or ?) <DOOR>: **<ENTER>**
Insertion point: **2800,1000**
X scale factor <1>/Corner/XYZ: **<ENTER>**
Y scale factor (default=X): **<ENTER>**
Rotation angle <0>: **90**

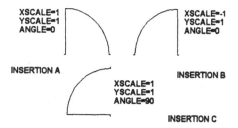

Figure 6.5 Inserting doors

Productivity tip : In many cases it is quicker to use the keyboard for block insertion rather than the dialog box. You will need to experiment with the different methods to see which one you prefer. This version of the command is also needed if you create macros or scripts for block insertions.

Global blocks

As mentioned above, blocks can be considered as mini-drawings. The converse is also true. Drawings themselves can be considered as blocks. Indeed a whole drawing can be inserted into the current drawing using the Insert dialog box in the normal way. Instead of giving picking the Block button, you pick **File** (Figure 6.4). You then select the drawing file name in the same way as if you were opening a drawing. The insertion base point will normally be the origin of the drawing but can be set to any point. Thus, all drawings are also blocks. As they are available to be inserted in any other drawing I call this type of block the "global block".

Blocks created in the same way as the DOOR above are only available within the drawing in which they were defined. They can be converted into global blocks, available to all drawings by using the **WBLOCK** command. This makes a copy of the block to a standard AutoCAD drawing file. WBLOCKs retain their layer, color and linetype settings. To write the DOOR block to a drawing file called "WDOOR.DWG" use the following command sequence.

The Create Drawing File dialog box will appear. Go to the "File name:" box, click the cursor and type **WDOOR**. Then click "Save" and supply the block name.

Command: **WBLOCK**
File name: **WDOOR** or use the dialog box
Block name: **DOOR**

The block is copied to WDOOR.DWG. The original is still intact in the current drawing. There are two other options at the "Block name" prompt above. If the block and the drawing to which it is to be saved have the same name, you type "=". If you type "*" in place of the block name the whole ofthe current drawing will be written to the globalblock.

Only layers that are actually used in the block are retained in WDOOR. In this case all the entities were drawn on layer 0. Normally when a WBLOCK or BLOCK is inserted it will be put on the current layer in the receiving drawing. However, it will still retain its own layer information for the entities making up the block. Thus if a block was originally created on the FITTINGS layer its entities will always be inserted on that layer. If the receiving drawing doesn't already have a layer called FITTINGS one will be created automatically. The one exception to this is a block created on layer 0. Such blocks and their component entities will always be inserted onto the receiving drawing's current layer.

Inserting a global block is much the same process as that described above. Use the INSERT command and give the WBLOCK name, ie "WDOOR", as the block name and proceed as before. You don't include the ".DWG" extension in the block name. As long as there is no block called "WDOOR" already in the drawing AutoCAD will search the current DOS directory for the drawing "WDOOR.DWG".

As a corollary to WBLOCK insertion, any AutoCAD drawing file can be INSERTed in another drawing. You simply give the full drawing name at the "Block name:" prompt or pick the File button in the Insert dialog box and carry on as usual. The insertion base point will be the origin of the external drawing unless a different base point has been specified. When creating a drawing to be used later as a block you can set a suitable base point with the BASE command. For example, to set the point (100,100) as the insertion base point of a drawing you would use the following command sequence:

Command: **BASE** Base point <0.0,0.0,0.0>: **100,100**

To set the base point back to the origin use the command once more.

Command: **BASE** Base point <100.0,100.0,0.0>: **0,0**

Making a library of useful symbols

The main items in our small modern fitted kitchen are the washbasin, the cooker, refrigerator or refrigerator-freezer, washing machine, dishwasher, storage units and worktops. There are of course many more items that could be included but those mentioned will suffice to illustrate the different block definition methods and also some new editing commands.

Before starting on the rest of the symbols, delete all the doors that you inserted above and change to the FITTINGS layer.

Command: **ERASE**
Select objects: **All**
3 found.
Select objects: **<ENTER>**
Command: **−LAYER** or pick the layer pull down list fron the toolbar
?/Make/Set/.../: **S**
New current layer <0>: **FITTINGS**
?/Make/Set/...: **<ENTER>**

Note that in the selection, 3 objects were found. Each block is considered by AutoCAD as a single object.

The kitchen sink

To draw the double drainer and basin shown in Figure 6.6 use the RECTAN-GLE command. A LINE with rectangular ARRAY can be used to draw the drainers. The curved corners of the basin are formed by filleting the rectangle. You might find it useful to zoom in using a window from (800,800) to approximately (2800,2200).

Command: **RECTANGLE**   or pick Draw/Rectangle from the menu
Chamfer/.../Fillet/...<First corner>: **1000,1000**   (A)
Other corner: **@1500,600**   (B)

The basin is another rectangle. To round the corners use the **Fillet** option in the prompt and give a radius of 50.

Command: **<ENTER>**
RECTANG
Chamfer/.../Fillet/...<First corner>: **F**
Fillet radius for rectangles <0>: **50**
Chamfer/.../Fillet/...<First corner>: **1500,1100**   (C)
Other corner: **@500,400**   (D)

There is a more general FILLET command which works on lines, polylines and other entities. To knock the sharp corners off the drainer (the larger rectangle), try the following. The first execution of Fillet is to set the radius for the rounded corners. The second is to apply it to the larger rectangle (Figure 6.6b).

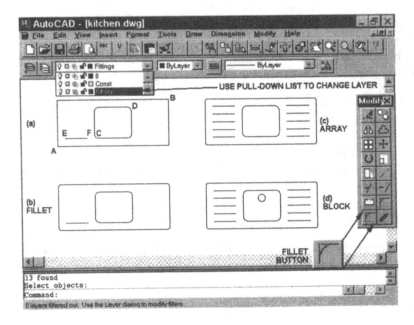

Figure 6.6 Filleting the basin

Command: **Fillet**   or pick **Modify/Fillet**
(TRIM mode) Current fillet radius = 10.0
Polyline/Radius/Trim <Select first object> **R**
Enter fillet radius <10.0>: **25**
Command: **<ENTER>**
Fillet
(TRIM mode) Current fillet radius = 10.0
Polyline/Radius/Trim <Select first object> **P**
Select 2D polyline: **1000,1000**   The large rectangle at (A)
4 lines were filleted

The drainer ribs are made up of parallel lines in a regular array.

Command: **LINE**
From point: **1100,1100**   (E)
To point: **@300,0**   (F)
To point: **<ENTER>**

The basin is taking shape and should look like Figure 6.6(b). The rest of
the drainer can be made by using a 5 row by 2 column array as shown in
Figure 6.6(c).

Command: **ARRAY**
Select objects: **LAST**   (This picks up the last line EF)
Select objects: **<ENTER>**
Rectangular or Polar array (R/P): **R**
Number of rows (---) <1>: **5**
Number of columns (| | |) <1>: **2**
Unit cell or distance between rows (---): **100**
Distance between columns (| | |): **1000**

Finally, add the drainage hole using a circle.

Command: **CIRCLE**
3P/2P/TTR/<Center point>: **1750,1400**
Diameter/<Radius>: **50** This completes the double drainer washbasin.

All that remains is to make it into a block. If you pick **Draw/Block/Make...**
you will get the same dialog box as in Figure 6.3. Give the block name as
BASIN with an insertion point of **1000** for both x and y and then pick
Select objects button. When everything has been selected, exit from the
"select objects" procedure by pressing **<ENTER>** once more.

Command: _bmake
Select objects: **Window**
First corner: **1000,1000**
Other corner: **@1500,600**
13 found.
Select objects: **<ENTER>**

Back at the dialog box, uncheck the Retain Objects box and pick **OK**.

Colored blocks

As blocks are inserted onto the layer names that they were created on, they
will take up the color of that layer. This means that if you change the color of
layer FITTINGS to blue and insert the BASIN block it will be drawn blue. In
many instances symbols or blocks have meaningful colors which we would like
to keep constant, irrespective of the color setting of the layer. In this section a
fridge-freezer block (Figure 6.7) will be defined to have a constant blue color
and a cooker block will be partly defined as red in the following section.

The color of an AutoCAD entity can be defined by its layer or may be
specially set using the COLOR command. Set the entity color to blue and
draw the fridge block. Either type the command or pick the color control box
at the left end of the toolbar and use the dialog box as in Chapter 3.

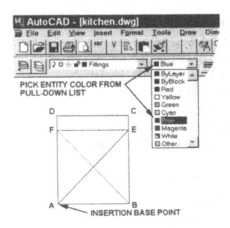

Figure 6.7 Fridge-freezer block

```
Command: COLOR
New object color: BLUE
Command: LINE
From point: 1000,1000                                    (A)
To point: @500,0                                         (B)
To point: @0,600                                         (C)
To point: @-500,0                                        (D)
To point: @0,-600                                        (A)
To point: @500,500                                       (E)
To point: @-500,0                                        (F)
To point: @500,-500                                      (B)
To point: <ENTER>
Command: COLOR
New entity color: BYLAYER
```

This completes the fridge-freezer in blue. Always reset the color back to "BY-LAYER" when finished with the special color. This means that new entities will be drawn in the default color assigned to the layer.

Now to group the entities into a block. You can use the BMAKE command as before or use the keyboard version of the command, **BLOCK**.

```
Command: BLOCK
Block name (or ?): FFREEZER
Insertion base point: 1000,1000
Select objects: Window
First corner: 1000,1000
```

Other corner: **@500,600**
7 found.
Select objects: **<ENTER>**

Editing a block

To draw the cooker in Figure 6.8 use the normal color for the outline and red
circles for the cooking elements. The array command can be used to copy the
hot-plate elements. For convenience you can edit the fridge-freezer block. To
do this you will have to insert the block and break it down into its constituent
entities. The EXPLODE command does just that.

Command: **INSERT**
Block name (or ?) <DOOR>: **FFREEZER**
Insertion point: **1000,1000**
X scale factor <1>/Corner/XYZ: **<ENTER>**
Y scale factor (default=X): **<ENTER>**
Rotation angle <0>: **<ENTER>**
Command: **EXPLODE** or select if from **Modify** pull-down menu.
Select objects: **LAST**
1 found
Select objects: **<ENTER>**

Now use the change command to set the colors to "bylayer" and then erase
the diagonal lines. Pick **Modify/Properties...** from the menu bar.

Command: _ai_propchk
Select objects: **900,900**
Other corner: **2000,2000**
7 found.
Select objects: **<ENTER>**
DDCHPROP loaded.

The Change properties dialog box should pop up. Pick the **Color** button which
brings the Select color dialog. Now pick **BYLAYER** and **OK**. Pick **OK** once
more to confirm the change of property and proceed to erase the diagonals.

Command: **ERASE**
Select objects: **1100,1100**   (One of the diagonal lines)
1 found
Select objects: **1400,1100**   (The other diagonal line)
1 found
Select objects: **<ENTER>**

Now set the color red and draw the cooking rings. We will first draw the bottom left ring as a circle. The OFFSET command will then be used to produce a series of concentric circles. Finally the ring will be arrayed to finish the cooker.

```
Command: COLOR
New entity color <BYLAYER>: RED
Command: CIRCLE
3P/2P/TTR/<Center point>: 1150,1150
Diameter/<Radius>: 75
```

It might help to zoom in close when doing the offsets. Then pick **Modify/Offset**.

```
Command:Zoom
All/.../Window/<Realtime>: 1000,1000
Other corner: 1300,1300
Command: REGEN                    (To regenerate the image)
Command: Offset
Offset distance or Through <1.0>: 10
Select object to offset: Pick a point on the circle A
Side to offset? Pick a point inside the circle
```

This creates the circle B in Figure 6.8. Now select the inner circle for the subsequent offset.

```
Select object to offset: Pick a point on the circle B
Side to offset? Pick a point inside the circle B
Select object to offset: Pick a point on the circle C
Side to offset? Pick a point inside the circle C
Select object to offset: Pick a point on the circle D
Side to offset? Pick a point inside the circle D
Select object to offset: <ENTER>
Command:Zoom
All/.../Window/<Realtime>: P
Command: ARRAY
Select objects: 1050,1050
Other corner: 1250,1250
5 found
Select objects: <ENTER>
Rectangular or Polar array (R/P): R
Number of rows (---) <1>: 2
Number of columns (| | |) <1>: 2
Unit cell or distance between rows (---): 200
```

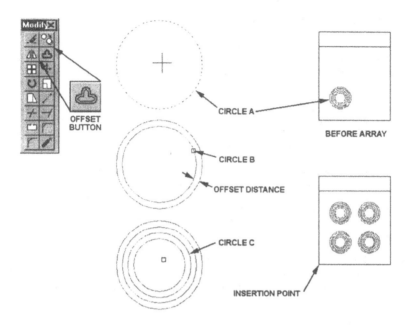

Figure 6.8 Cooker block

Distance between columns (| | |): **200**

This should now look like the picture in Figure 6.8. The only way to edit part of a block is to decompose it completely into the original entities. One way of achieving this is to EXPLODE it as above. A similar effect results by prefixing the block name with an asterisk when inserting it. For example:

Command: INSERT
Block name (or ?) <DOOR>: *FFREEZER
Insertion point: 1000,1000
Scale factor <1>: <ENTER>
Rotation angle <0>: <ENTER>

This gives the same result as the INSERT and EXPLODE used above. This method can only be used when actually inserting a block. EXPLODE can be used on a block at any time after it has been inserted. There is also an EXPLODE option in the Insert dialog box in the lower left corner (Figure 6.4).

Now to block the cooker and reset the color to bylayer.

Command: **BLOCK**
Block name (or ?): **COOKER**

```
Insertion base point: 1000,1000
Select objects: Window
First corner: @
Other corner: @500,600
25 found.
Select objects: <ENTER>
Command: COLOR
New entity color <1 (red)>: BYLAYER
```

One must be careful when defining special colors for blocks or parts of blocks. As blocks can be nested within other blocks it can become difficult to keep track of the colors. The best policy is to keep the coloring of blocks as simple as possible and to use special color settings sparingly.

There is one other color setting that has not been discussed so far. For blocks that are made up of entities on different layers, the BYBLOCK color option can be used. This forces all the entities making up the block to have the same color (whatever color the block is set to). This is irrespective of the color settings of the individual layer and is the converse of having a multi-color block on one layer.

Finally, all of the rules given above for the COLOR command can be equally applied to the LINETYPE command. LINETYPE can be used to override the Ltype settings on individual layers. It works in exactly the same way as COLOR.

Assigning text information to blocks

One of the strongest reasons for using blocks in AutoCAD drawings is that extra non-graphic information can be assigned to the blocks. Furthermore, the attribute information attached to a block can be varied each time the block is inserted. All this information can be gathered together and written to an external file which can then be transferred to a bill of materials program to extract quantities and cost estimates.

In this section a simple attribute will be defined for an electrical appliance block. This will identify what type of appliance has been inserted. Attributes are also defined for cupboard units and worktop finishes.

Defining an attribute

An attribute is treated by AutoCAD like any other drawing entity. Once it has been defined it can be included with other entities to form a block. To create the electrical appliance block (Figure 6.10), first draw its outline. Use

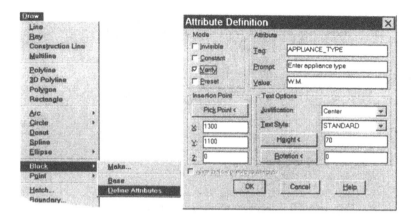

Figure 6.9 Attribute menu and dialog box

the RECTANG command. We will have to set the fillet radius back to zero since the last time we used RECTANG we put the radius at 50.

Command: **RECTANG**
Chamfer/.../Fillet/.../<First corner>: **Fillet**
Fillet radius for rectangles <50>: **0**   (For sharp corners)
Chamfer/.../Fillet/.../<First corner>: **1000,1000**
To point: **@600,500**

Now include hot and cold inlet pipes and the waste outlet point at the back of the box.

Command: **LINE**
From point: **1100,1500**
To point: **@0,100**
To point: **<ENTER>**
Command: **<ENTER>**
LINE From point: **1200,1500**
To point: **@0,100**
To point: **<ENTER>**
Command: **<ENTER>**
LINE From point: **1300,1500**
To point: **@0,100**
To point: **<ENTER>**

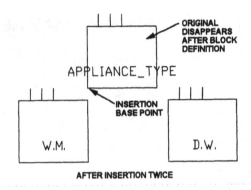

Figure 6.10 Electrical appliance blocks with attributes

Now to define the attribute pick **Draw/Block** from the menu bar and then pick **Define Attributes...**, as shown in Figure 6.9. This brings up the Attribute Definition dialog box.

The dialog box is divided into four sections, namely, Mode, Insertion point, Attribute and Text options.

In the Mode part of the dialog box click the "Verify" box to set this mode on. This means that when the block is inserted you will be asked to verify that the value of the attribute is correct. All the other modes are okay for this block. It will not be invisible, constant, or preset but will be verified.

The insertion point **1300,1100** is in the lower middle of the rectangle already drawn. If you pick the Pick Point button you can use the mouse to select the point from the graphics screen.

Moving to the Attribute part of the dialog box, the attribute tag is the name for that attribute and will be used by AutoCAD to identify it. Any characters can be used for the tag except blank spaces. Type **APPLI-ANCE_TYPE** in this field. The prompt is the message that will be displayed when the block is being inserted. It should be clear so that others using your library of blocks can understand what is required of them. The default value will be offered in AutoCAD's usual way. The "W.M." stands for washing machine; the alternative will be "D.W." for dishwasher. You can use more descriptive attribute values if you wish. The command then asks for the start point of the attribute text in a similar fashion to the TEXT command. Position it centrally in the appliance box and use a large text height to make it visible.

Finally, you must specify how the attribute will be displayed. The text options are similar to those for the standard DTEXT command. Pick the arrow at the end of the Justification field. This gives a pull-down list of the different types of justification. Pick **Center**. Then change the text height to **70**. When your dialog box matches that in Figure 6.9 pick **OK**.

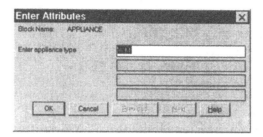

Figure 6.11 Enter Attributes dialog box

The text "APPLIANCE_TYPE" should be written across the box as shown in Figure 6.10. This will be replaced by the actual attribute value when the block has been created and inserted. To create the block with the attribute, pick **Draw/Block/Make...** or type **BLOCK** at the command prompt. Select the objects for inclusion using a window big enough to surround the attribute as well.

Command: **BLOCK**
Block name (or ?): **APPLIANCE**
Insertion base point: **1000,1000**
Select objects: **300,700**
Other corner: **2400,1900**
5 found.
Select objects: **<ENTER>**

Now to check that it works insert it at (1000,1000). If you are confronted with the Attribute Dialog Box, similar to the one shown in Figure 6.11, just pick the **OK** box to proceed. To disable the dialog box reset the system variable, ATTDIA, to 0 by typing the variable name at the command prompt followed by the desired value (ATTDIA = 1 enables the dialog box). You can decide whether you prefer entering the attributes via the dialog box or the keyboard. If a block has a number of attributes then the dialog box is quite useful. If the block has only one or two attributes then the keyboard is quicker. Note that the verify mode only applies to keyboard input.

Command: **ATTDIA**
New value for ATTDIA <>: **0**
Command: **INSERT**
Block name (or ?) <FFREEZER>: **APPLIANCE**
Insertion point: **1000,1000**
X scale factor <1>/Corner/XYZ: **<ENTER>**

Y scale factor (default=X): **<ENTER>**
Rotation angle <0>: **<ENTER>**
Enter attribute values
Enter appliance type <W.M.>: **<ENTER>**
Verify attribute values
Enter appliance type <W.M.>: **<ENTER>**

The block should now be inserted with the "W.M." written in it. Try inserting it again but give "D.W." as the attribute value. You should get something like Figure 6.10. Using attributes in this way saves having to define different blocks for each type of large electrical appliance.

The attributes of a block can be visible or invisible (See the Mode section in Figure 6.9). This is useful when the text is not relevant to the actual picture or if it would crowd the drawing too much. In the following sequence you will define a block for the cupboard units with an invisible attribute giving information on the number of doors on the unit. As blocks can have more than one attribute you can also include an attribute for the type of finish the customer requires on the unit.

The standard unit size in Figure 6.13 is 500mm wide by 600mm deep. Two units can be joined to form a double, etc. Draw the unit using a polyline this time and use a thicker line to indicate the side with the door. Be sure to **ERASE** all the copies of the block you have just inserted.

Command: **PLINE**
From point: **1000,1000**
Current line-width is 0.0000
Arc/Close/.../<Endpoint of line>: **@600<90**
Arc/Close/.../<Endpoint of line>: **@500,0**
Arc/Close/.../<Endpoint of line>: **@0,-600**
Arc/.../Width/<Endpoint of line>: **Width**
Starting width <0.0000>: **25**
Ending width <25.0000>: **<ENTER>**
Arc/Close/.../<Endpoint of line>: **CLOSE**

Now define the first attribute for the number of doors in the unit. Pick **Draw/Block/Define Attributes...** from the menu bar. Then click **Invisible** and **Verify**. The insertion point is **(1250,1400)**. Type **DOORS** for the attribute tag, the prompt as shown and give a default value of **1**. The text options are as before. When your display matches Figure 6.12 pick **OK**.

Now repeat this for the surface finish attribute. An alternative method is to use the ATTDEF command rather than the dialog box.

Command: **ATTDEF**
Attribute modes – Invisible:Y Constant:N Verify:Y Preset:N

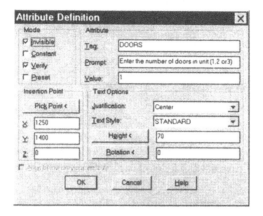

Figure 6.12 Cupboard attributes

Enter (ICVP) to change, RETURN when done: **<ENTER>**
Attribute tag: **FINISH**
Attribute prompt: **Enter the type of surface finish**
Default attribute value: **OAK**
Justify/Style <Start point>: **J**
Align/Center/...: **C**
Center point: **1250,1100**
Height <>: **70**
Rotation angle <0>: **<ENTER>**

You are ready to make the block now.

Command: **BLOCK**
Block name (or ?): **CUPBOARD**
Insertion base point: **1000,1000**
Select objects: **Window**
First corner: **300,300**
Other corner: **1900,1900**
3 found.
Select objects: **<ENTER>**

When you insert this block later you will be prompted for the two attributes.
The order will be the reverse of the definition, ie you will be prompted for
the surface finish first. The block will be drawn with both attributes invisible
(Figure 6.13). You can experiment with inserting this block or wait for the
next drawing in which all the blocks will be used to assemble the AutoCAD
Express Fitted Kitchen. Be sure to **ERASE** everything before proceeding.

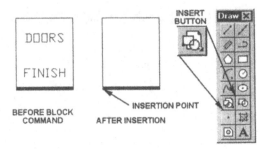

Figure 6.13 The cupboard with two attributes

To get a listing of all the blocks in the drawing you can use the "?" option with either the BLOCK or INSERT commands. The display will flip to text mode and give the names. Alternatively, pick **Insert/Block...** from the menu bar. Then pick the **BLOCK** button in the Insert dialog box. You can then pick the desired block from the list by double clicking it followed by **OK** (Figure 6.14).

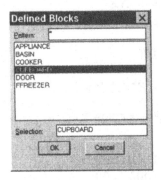

Figure 6.14 List of defined blocks

Drawing the kitchen

Before inserting all the blocks, let's do the outline of the kitchen. It's only a small kitchen with a simple shape. This should be done on layer 0. You can see from Figures 6.15 and 6.17, the type of thing required. To represent the wall, a pair of line 100 units apart are drawn. These could be drawn using PLINE

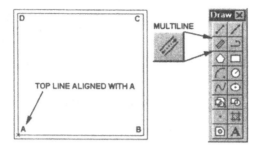

Figure 6.15 Multiline walls

and then Offset. However, since Release 14 there is a more elegant solution using Multiline and Multiline edit.

Command: −**LAYER** or use the tool bar
?/Make/Set/...: **S**
New current layer <FITTINGS>: **0**
?/Make/Set/...: **<ENTER>**
Command: **Zoom**
All/.../Window/<Realtime>: **ALL**
Command: **MLINE**
Justification = Top, Scale = 20.0, Style = STANDARD
Justification/Scale/STyle/<From point>: **Scale**
Set Mline scale <20.0> **100**   (The distance between lines)
Justification/Scale/STyle/<From point>: **1000,1000**   (A)
<To point>: **@3300,0**   (B)
Undo/ <To point>: **@0,3100**   (C)
Close/Undo/ <To point>: **@-3300,0**   (D)
Close/Undo/ <To point>: **CLOSE**

Note that AutoCAD takes care of the corners and that the Justification means that the points input are the corners of the inner lines. The first point was on the top multiline. The other justifications are bottom or zero. The latter positions the lines equidistant from the selected points

The multiline drawn here uses the STANDARD style. You can create elaborate multiline styles by picking **Format/Multiline Style**.... This executes the MSTYLE. With MSTYLE you can specify how many lines are to be drawn, their colors and linetypes. Since the STANDARD is the most useful I will leave you to explore MSTYLE via the AutoCAD Help file and move on with the kitchen plan.

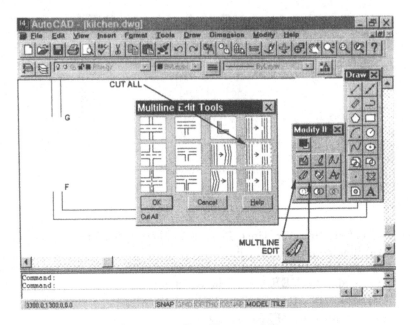

Figure 6.16 Breaking a multiline

Now that the walls are up we must break through some door openings. This introduces AutoCAD's multiline editing command, MLEDIT. Before this, Zoom in for a closer look.

Command: **ZOOM**
All/.../Window/<Scale(X/XP)>: **W**
First corner: **500,500**
Other corner: **3000,2400**

The multiline edit button is on the **Modify II** toolbar. Use **View/Toolbars...** to display Modify II (Figure 6.16. Executing MLEDIT brings the pop up selection box of edit tools shown in Figure 6.16. Pick the "Cut All" icon shown on the right of the middle row followed by **OK**.

Command: **MLEDIT**
Select mline: **1000,1200**   (F)
Select second point: **@0,800**   (G)
Select mline (or Undo): **<ENTER>**
Command: **ZOOM**
All/.../Window/<Scale(X/XP)>: **Previous**

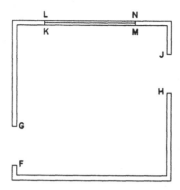

Figure 6.17 The empty kitchen

This breaks an 800mm opening from point F to G (Figure 6.16). Now repeat the command to break through for the second door, HJ, on Figure 6.17.

> Command: **MLEDIT** and select the **Cut All** icon and **OK**
> Select mline: **4300,2700**   (H)
> Select second point: **@0,800**   (J)
> Select mline(or Undo): **<ENTER>**

Both MLEDITs could have been done within the one operation as the command prompts for further mlines to cut. The final stage in making the opening is to draw the short lines to close off the open ends of the walls.

> Command: **LINE**
> From point: **1000,1200**   (F)
> To point: **@−100,0**
> To point: **<ENTER>**
> Command: **<ENTER>**
> LINE
> From point: **1000,2000**   (G)
> To point: **@−100,0**
> To point: **<ENTER>**

And now for the other opening.

> Command: **LINE**
> From point: **4300,2700**   (H)
> To point: **@100,0**
> To point: **<ENTER>**
> Command: **<ENTER>**

```
LINE
From point: 4300,3500                                              (J)
To point: @100,0
To point: <ENTER>
```

The window, KLMN, is drawn with two short lines and a horizontal glazing line.

```
Command: <ENTER>
LINE
From point: 1600,4100                                             (K)
To point: @0,100                                                  (L)
To point: <ENTER>
Command: <ENTER>
LINE
From point: 3600,4100                                            (M)
To point: @0,100                                                 (N)
To point: <ENTER>
```

Now draw the window glazing as a horizontal line.

```
Command: LINE
From point: 1600,4150                              (Mid pt of line KL.)
To point: @2000,0                                  (Mid pt of line MN.)
To point: <ENTER>
```

Assembling the fitted kitchen

Assembling the fittings and fixtures of the kitchen is now just a matter of inserting all the blocks in their correct locations. If the customer wants things moved around then that's no problem with AutoCAD. You can remain on Layer 0 since blocks are inserted onto their original creation layer.

Insert the basin first with a view out the window. Pick **Insert** from the menu bar, then pick the **Block...** button followed by **BASIN**. Complete the dialog box as shown in Figure 6.18. Disable the "Specify Parameters on Screen" and input the coordinates of the insertion point, P, (**1600,3500**) as shown on Figure 6.18. Use the default scales and pick **OK**.

Now put the washing machine and dishwasher beside the basin (Figure 6.18). In the following sequence, ATTDIA is 0 and keyboard input is used. You can use the dialog box if you wish. The first appliance is located by typing the coordinates while the second uses the Object snap intersection point at R.

```
Command: INSERT
```

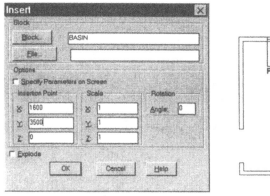 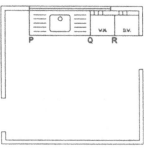

Figure 6.18 Inserting the Basin and appliances

Block name (or ?) <BASIN>: **APPLIANCE**
Insertion point: **3100,3500**   (Q)
X scale factor <1>/Corner/XYZ: **<ENTER>**
Y scale factor (default=X): **<ENTER>**
Rotation angle <0>: **<ENTER>**
Enter appliance type <W.M.>: **<ENTER>**
Verify attribute values
Enter appliance type <W.M.>: **<ENTER>**
Command: **INSERT**
Block name (or ?) <APPLIANCE>: **<ENTER>**
Insertion point: **INT** of   Pick Q or type **3700,3500**
X scale factor <1>/Corner/XYZ: **<ENTER>**
Y scale factor (default=X): **<ENTER>**
Rotation angle <0>: **<ENTER>**
Enter appliance type <W.M.>: **D.W.**
Verify attribute values
Enter appliance type <D.W.>: **<ENTER>** The rest of the block infor-

mation is given in tabular form. Use the INSERT command and the appropriate responses taken from Table 6.2. Figure 6.19a shows the fitted kitchen.

Editing attributes

The first thing to do now is to check that all the attributes on the drawing
are in fact correct. To make the invisible attributes appear on the drawing
use the **ATTDISP** command. This command allows you to alter the display

Table 6.2 Block insertion parameters

Block name	Insertion point	X scale	Y scale	Rotation	Attribute values
FFREEZER	3700,2600	1	1	−90	
COOKER	1600,2600	1	1	90	
CUPBOARD	1600,3100	2	1	90	PINE, 1
CUPBOARD	1600,2100	1	1	90	PINE, 1
CUPBOARD	3700,2100	2.2	1	−90	PINE, 2
CUPBOARD	3700,1600	3	1	180	PINE, 3
DOOR	950,2000	1	1	−90	
DOOR	4350,2700	−1	1	−90	

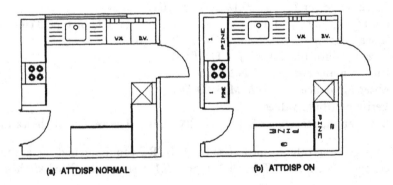

(a) ATTDISP NORMAL   (b) ATTDISP ON

Figure 6.19 The fitted kitched

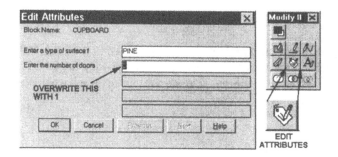

Figure 6.20 Dialog box for attribute edits

setting for all the attributes in the drawing. The normal display mode is that only attributes defined as visible are shown. Setting ATTDISP to ON causes all attributes to become visible irrespective of their definition. ATTDISP OFF would cause all to become invisible. The setting can be altered from the menu bar by picking **View/Display** followed by **Attribute Display**.

Command: **ATTDISP**
Normal/ON/OFF <Normal>: **ON**

From Figure 6.19 we can see that the cupboard in the lower right-hand corner has 2 doors. Since it is in the corner, one of these doors will be blocked by the other cupboard. To save the customer unnecessary expense only one door should be provided. Thus the attribute must be edited. The easiest way to do this is to use the **Edit Attribute** button on the Modify II toolbar. This executes the command **DDATTE** which prompts you to select a block from the display and shows the current values of all attributes in the dialog box in Figure 6.20.

Command: _ddatte
<Select block>/Undo: **3700,1900**

The dialog box shown in Figure 6.20 should then appear. This gives the two attribute values and their prompt. To change the 2 doors to 1 move the arrow cursor and click in the rectangle containing the 2. Then type 1. This should overwrite the old value. You may have to use the backspace key to delete the "2" before typing the 1. To execute this change pick the **OK** box at the bottom of the dialog screen. When the dialog box disappears the "2" in the block will be changed to "1".

To set the attribute display back to normal use ATTDISP once more.

Command: **ATTDISP**

Normal/ON/OFF <ON>: **Normal**

The ATTDISP setting does not effect the ability to edit the attributes. You can edit attributes in exactly the same way when they are invisible. However, it is not always easy to notice that an edit is required when the values are hidden.

A simple bill of materials

Information about all blocks that have attributes can be extracted from the drawing and output to a text file. The text file could then be incorporated in a report, a spreadsheet or a bill of quantities. There are three ways to extract this information, namely, using AutoCAD's DXF format, a CDF file or an SDF. The most useful way is to use a template file to control which attributes are required for output to either an SDF or a CDF file.

A CDF file (Comma Delimited File) is an ASCII text file where the block's information fields are separated by commas. In an SDF they are separated by spaces while the DXF file uses AutoCAD's drawing exchange protocol. The latter is of use to program developers but is rather cumbersome and generated large data files.

In order to use the SDF or CDF method you first have to define a template file. This has to be an ASCII file created with a text editor such as Notepad or Wordpad. The file name must have the extension ".TXT". The file given below is suitable for use on the KITCHEN drawing. The comments in brackets should *not* be included in the actual EXKITCH.TXT file (Table 6.3). The first item on each line is a key-word indicating what is to be extracted, the second gives the format for writing that information to the CDF or SDF file. If the information is a number then the second field should start with "N"; if it is text then "C" should be used. The first three digits after the "C" or "N" indicate how many spaces are to be reserved for that field. The second three indicate how many digits are required after the decimal point. Thus "N007001" would output a number such as "1234.5" but not "1234.56". Always leave one space for a possible minus sign. Make sure that there are no blank lines in the template file.

When you are in the drawing editor with the KITCHEN drawing you can generate the bill of materials using the DDATTEXT command. This does not appear on any of the standard menus, so you have to know about it. The dialog box shown in Figure 6.21 will then appear. Click the **SDF** format and give the template and output files as shown and pick the **Select objects** button.

Command:**DDATTEXT**
Select objects: **ALL**   (Non block objects will be ignored)

Table 6.3 EXKITCH.TXT

BL:NAME	C010000	(Block name with up to 10 characters)
FINISH	C007000	(Finish attribute with up to 7 characters)
APPLIANCE_TYPE	C007000	(Appl. type attribute with up to 7 characters)
DOORS	N006000	(No. of doors attribute up to 6 digit integer)
BL:XSCALE	N007001	(X scale factor, number with 1 digit after ".")
BL:YSCALE	N007001	(Y scale factor, number with 1 digit after ".")

Figure 6.21 Attribute extraction dialog box

Table 6.4 KITCHEN.TXT

APPLIANCE		W.M.	0	1.0	1.0
APPLIANCE		D.W.	0	1.0	1.0
CUPBOARD	pine		1	1.0	1.0
CUPBOARD	pine		1	2.0	1.0
CUPBOARD	pine		1	2.2	1.0
CUPBOARD	pine		3	3.0	1.0

28 found
Select objects: <**ENTER**>

AutoCAD should echo "6 records in extracted file" when the **OK** button is picked. Now to see the result use Microsoft Wordpad or another editor to open the file. It should look something like Table 6.4.

The order may not be the same in your file and some text editors handle the end of line characters differently requiring the user to do some formatting. The above file should look ok in MS Wordpad.

Only blocks that contain attributes included in the template file are extracted. The DOOR, BASIN COOKER and FFREEZER block don't appear as they have none of the attributes. You can include more information fields, if desired, to extract the insertion points and other data. See the AutoCAD Customization Guide for the full list of key-words.

Hints on using blocks

Blocks are good for commonly used shapes and symbols. You can build up a library of symbols using WBLOCK. This is more efficient than using a template drawing that contains many block definitions.

Try to develop a consistent method for naming your blocks and store the WBLOCKS in a separate disk folder. For example you might create a subfolder of \ACAD called "\ACAD \SYMBOLS" and include that path in the BLOCK file name when inserting. The more symbols you have the more effort you will have to devote to managing this storage.

Blocks forming part of a symbols library should always contain some attribute definitions. The minimum required would be a constant attribute for something like a part number. The user would not be prompted for a value for such an attribute. This will enable information about them to be extracted for bill of material purposes.

If a complicated object (say, containing more than 20 entities) appears more than once in a drawing then BLOCK it and INSERT it as required. This is more efficient than simply copying all the entities, as AutoCAD stores the shape of the block only once. The only other items of information required are the relevant insertion points. This saves memory and disk space.

Avoid using nested blocks if possible. By nested, one means that one block can contain others. This can lead to difficulties if you want to make changes to the block definition at a later date.

AutoCAD's inquiry commands

A number of other information extraction commands are available in the Tools/Inquiry pull-down menu. These allow you to get the area enclosed by a polyline (**AREA** command), the distance between two points (**DIST** command) or the coordinates of a particular point (**ID**). **LIST** allows you to find out all the information about any entities. This is helpful for finding out which layer an entity has been drawn on. The **STATUS** command gives a listing of the current drawing status and also the amount of memory available in the computer and on the disk. This is useful for keeping tabs on your hardware and the current AutoCAD default settings.

To finish this session pick **File/Exit** followed by **Save Changes** if you are prompted to do so.

Summary

This chapter has introduced the fundamentals of AutoCAD block creation. Blocks are a useful feature for storing standard symbols. They can be exported from the original drawing and so be made available to other drawings. Some more editing features have been used to create the blocks. Finally, the information extraction facility has been demonstrated.

You should now be able to:

Create blocks and global blocks.
List the names of blocks in a drawing.
Assign text information to blocks and edit it.
Round off corners with the FILLET option.
Draw multiple lines
Offset objects and break gaps in multilines.
Alter the entity color.
Interrogate the drawing database.

Chapter 7 ADVANCED DRAWING AND DIMENSIONING

General

Most of AutoCAD's drawing commands have been covered by now. In this chapter you will discover the reason for creating AutoCAD drawings at full scale. The various dimensioning and measuring commands all calculate their distances in the drawing units. To demonstrate the more important automatic dimensioning facilities you will draw two relatively simple objects, a mechanical engineer's gland and a chamfered pentagon (Figure 7.1). In doing this a few more new commands will be introduced. Then the dimensions will be added.

A template for drawing a gland

Start a new AutoCAD drawing using a template. In the Create New Drawing dialog box, pick **Use a Template** button and scroll down thelist to pick **Acadiso.dwt** as shown in Figure 7.2. Pick **OK** in the dialog box. The template file is equivalent to the prototype file in earlier releases of AutoCAD. It contains a number of settings which will be helpful when we add dimensions.

You can save any drawing as a template by picking Save As and selecting the file type as "dwt". The resulting template then becomes the starting point for future drawings when selected.

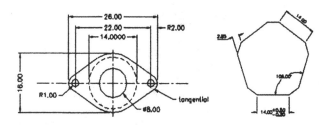

Figure 7.1 Engineer's gland

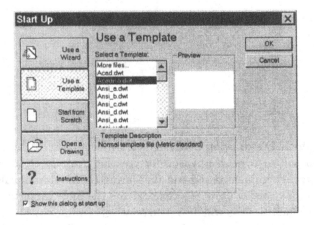

Figure 7.2 Drawing template

Table 7.1 Layer settings for drawing GLAND

Layer name	State	Colour	Linetype
0	On	7 (white)	CONTINUOUS
CLINE	On	7 (white)	CENTER
DASH	On	7 (white)	DASHED
DIMENSIONS	On	7 (white)	CONTINUOUS
GLAND	On	7 (white)	CONTINUOUS
POLYGON	On	7 (white)	CONTINUOUS

Current layer: CLINE

Give the drawing the name "GLAND" by picking **File/Save**. Set the LIMITS from (0,0) to (65,45) and make sure that decimal units are being used with 4 places after the decimal point. You will need the layers given in Table 7.1. To begin the drawing put the center-lines of the circles at convenient locations (Figure 7.3) on the layer "CLINE". Double click the **Ortho** button on the status line to turn on the ORTHO mode. You may also find it helpful to set SNAP to 1 and GRID to 5. If the lines don't appear with dashes change the LTSCALE value to 0.1.

```
Command: LINE
From point: 13,20                                        (A)
To point: @44,0                                          (B)
To point: <ENTER>
Command: <ENTER>
LINE From point: 35,11                                   (C)
To point: @0,18                                          (D)
To point: <ENTER>
```

POINT and DIVIDE

These are the major axes for the gland's cylinder. The line AB is 44cm long and the distance from the center of the gland to the bolt holes on either side is 11cm. Thus the quarter points of AB can be used to position the holes. To find the quarter points use the **Draw/Point** from the menu bar followed by **Divide** (Figure 7.3). This can be used to divide a line, arc or polyline into any number of equal segments. Rather than actually break the line into different entities the DIVIDE command inserts AutoCAD POINTs at the relevant intervals. Here we wish to divide the line AB into 4 equal lengths.

```
Command: divide
Select object to divide: 16,20                    (Point on AB)
<Number of segments>/Block: 4
```

Unless you were very lucky, your dividing points will not be visible just yet. This will be fixed very shortly.

The "Block" option it the sequence above allows you to insert a named block at the dividing locations instead of points. The points should now appear along the line as shown in Figure 7.3. Another feature of this command is that all the points are put in the "previous" selection set and can be deleted by picking "ERASE" followed by "Previous".

POINTs are drawing entities. Their main use is for marking special locations for object snapping. Object snap "node" jumps to the nearest point

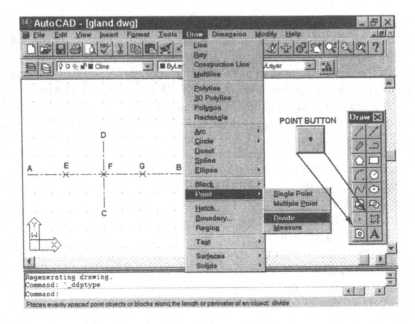

Figure 7.3 Dividing line

entity. These normally appear on the drawing as dots, which are not noticeable against the line **AB**. Since the dots have no dimension they are difficult
to see, particularly if the grid is on. AutoCAD gives a number of options for
the display of points in the Points Style dialog box (Figure 7.4). Pick **For-
mat/Point Style...** see the different styles. Then pick the fourth box on the
top row, containing "X". Then set the point size to 5% relative to the screen
size. This relative sizing means that no matter how far in or out we zoom and
magnify the drawing, the points will appear the same size. After picking **OK**
type **REGEN** to show the points in their new style.

AutoCAD's MEASURE command is very similar to DIVIDE but it puts
the point markers at multiples of a specified distance from the end point of the
entity. Thus you could use **MEASURE** to do the same as the above by giving
the distance to measure out as 11 units.

Command: MEASURE
Select object to measure: 16,20   (Point on AB)
<Segment length>/Block: 11

This will include as many markers as will fit on the line. With **MEASURE**
four will be drawn at 11, 22, 33 and 44 units from the point A. It is important
to pick the line to be measured near to the end you want the measurement

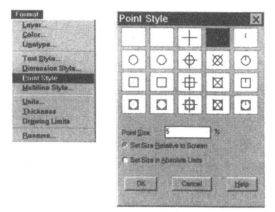

Figure 7.4 Point style dialog box

to start, particularly if the line length is not a whole multiple of the segment length.

You can now set an Object Snap Running mode to "Node" to snap the circle centers to the dividing points. Pick **Tools/Object Snap Settings**... and then pick **Node** in the dialog box shown in Figure 7.5. Then pick **OK**. Before drawing the circles, change to layer GLAND and draw circles at the points E, F and G (Figure 7.3).

Command: DDOSNAP
Command: **−LAYER** or use the tool bar.
?/Make/Set/...: **S**
New current layer <CLINE>: **GLAND**
?/Make/Set/...: **<ENTER>**

For the circles on Figure 7.6, use radii of 8, 7, 4, 2 and 1 as given below. Pick **Circle** button from the Draw toolbar and draw three large circles (Figure 7.6). The centers will automatically snap to the point entities. You can leave the cursor on the point F for the 3 circles.

Command: CIRCLE
3P/2P/TTR/<Center point>: (Pick point at F)
Diameter/<Radius>: **8**
Command: **<ENTER>**
CIRCLE
3P/2P/TTR/<Center point>: (Pick point at F)
Diameter/<Radius><8.0000>: **7**
Command: **<ENTER>**

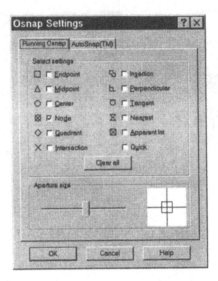

Figure 7.5 Object snap running mode

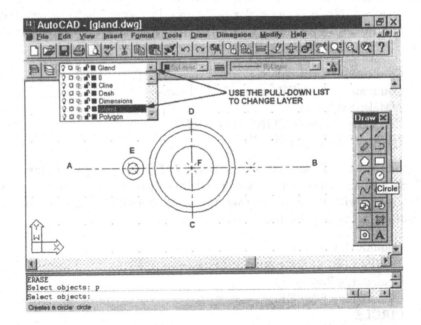

Figure 7.6 Gland circles

CIRCLE
3P/2P/TTR/<Center point>: (Pick point at F)
Diameter/<Radius>: **4**

Now draw the two small circles at the left-hand node. These have radii of 2 and 1 respectively.

Command: **CIRCLE**
3P/2P/TTR/<Center point>: (Pick point at E)
Diameter/<Radius>: **2**
Command: **<ENTER>**
CIRCLE
3P/2P/TTR/<Center point>: (Pick point at E)
Diameter/<Radius>: **1**

You will place the objects at G later using a mirroring operation. Now erase the three dividing points and turn off the Object Snap running mode. A quick way to do this is with the OSNAP command.

Command: **ERASE**
Select objects: **Previous**
3 found.
Select objects: **<ENTER>**
Command: **OSNAP**   or pick Tools/Object Snap Settings

Pick **Clear all** from the Osnap Settings dialog followed by **OK**.

If the points are not deleted and you get the message "No previous selection set" it is probably because you have used a command that required you to "Select objects:". If you do such a command after the DIVIDE and before the erase then the selection set will be altered. In that case you will have to delete the points individually.

You can build up a selection set of objects for use with the "previous" option in ERASE and other commands. This is done with the SELECT command. The format is just like ERASE but without anything being deleted.

Command: **SELECT**
Select objects: Pick objects or use Window, Crossing etc.
Select objects: **<ENTER>**

To draw the flange for the gland, zoom in on the five circles and draw two lines tangential to both of the outer circles (Figure 7.7). Use **ZOOM Realtime** to find a suitable magnification. The Zoom Realtime button is the magnifying glass with the plus and minus.

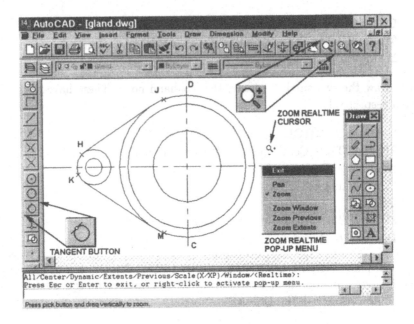

Figure 7.7 Tangential lines

Command: **ZOOM**
All/Center/.../<Realtime>: <**ENTER**>
Press Esc or Enter to exit, or right click to activate pop-up menu.

The cursor changes to a magnifying glass. Hold the left button down on the
mouse and move the cursor up the screen for increased magnification or down
for zooming out. Press the right button to bring the pop-up menu. Do this
and pick **Exit** when the screen is similar to Figure 7.7.
  To draw the lines use **Tools/Object Snap Settings...** with running
mode **Tangent**.

Command: **OSNAP** or pick Tools/Object Snap Settings...
Object snap modes: **TANGENT** or pick from dialog box.

AutoCAD will look for a tangent point every time a point is picked. As you
move the cursor over the circles you should get a pop-up comment of "Deferred
Tangent" from the AutoSNAP feature.

Command: **LINE**
From point: (Pick the small outer circle near point **H**)
To point: (Pick the largest circle near point **J**)
To point: <**ENTER**>

Note that the Object Snap mode overrides the ORTHO mode. Also note that the line did not appear until the second point was picked. This was because AutoCAD had to calculate the tangent point on the first circle and this was dependent on the second point of the line. Now repeat this process for points K and M.

Command: <ENTER>
LINE From point: (Pick the small outer circle near point K)
To point: (Pick the largest circle near point M)
To point: <ENTER>

Before going any further you must clear the running **OSNAP** or object snap mode to **None**. It is easy to forget about the object snap mode and that could lead to undesirable results when picking points later. The method below uses the keyboard version of the command. The minus causes AutoCAD to use the command as in earlier Releases of the program.

Command: **−OSNAP**   This executes a keyboard version.
Object snap modes: **NONE**

Trimming entities

The two outer circles must now be trimmed back to their intersection points with the tangents (Figure 7.8). The **TRIM** command works like EXTEND and can be found in the Modify menu. There is also a Trim button in the Modify toolbar. With this command, you are first prompted for the boundary lines or arcs, etc, to define the trimming edges. Then you specify the entities to trim.

Command: **TRIM**
Select cutting edge(s): (Projmode = UCS, Edgemode = No extend)
Select objects: (Pick the line HJ near its mid point)
Select objects: (Pick the line KM near its mid point)

The two lines should now appear ghosted. If anything else has been selected by mistake type **Remove** and pick the unwanted objects. When the selection of the boundaries is completed press <ENTER> to proceed with trimming the circles.

Select objects: <ENTER>
<Select object to trim>/Project/Egde/Undo: (Pick small circle at N)
<Select object to trim>/Project/Egde/Undo: (Pick large circle at P)
<Select object to trim>/Project/Egde/Undo: <ENTER>

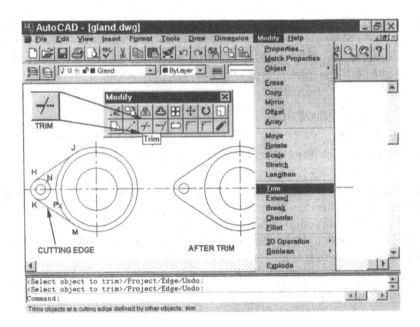

Figure 7.8 Trimming the flange

Now to complete this view you can mirror the half-flange about the center-line, CD. Pick **Modify/Mirror** from the menu bar.

Command: **MIRROR**
Select objects: **Window**
First corner: **21,12**
Other corner: **37,29**
4 found.
Select objects: **<ENTER>**
First point of mirror line: **35,11**   (C)
Second point: (Pick a point vertically above C using ORTHO)
Delete old objects? <N>: **<ENTER>**

The large circle can be trimmed on the other side as before.

Command: **TRIM**
Select cutting edge(s)...
Select objects: **44,24**   (Q)
1 found.
Select objects: **44,16**   (R)
1 found.

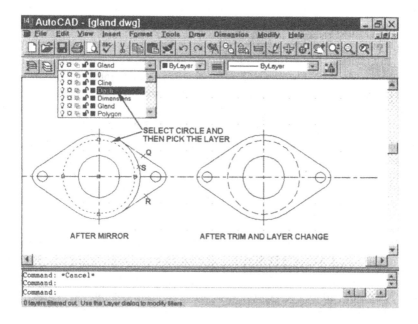

Figure 7.9 Plan view of gland

Select objects: **<ENTER>**
Select objects to trim: **43,19**   (S)
Select objects to trim: **<ENTER>**

For the plan view to be complete the 14cm diameter circle should be drawn in dashed linetype (Figure 7.9). To do this change it to the DASH layer. Pick the circle at T. This makes it ghosted and displays its grips. Now go to the Layer pull-down list and select **Dash**. This swaps the circle from the current layer over to Dash. If that didn't work, you can use the **Modify/Properties...** command.

Dimensioning

As the gland has been drawn to full scale, all the correct length information is already stored in the drawing. To extract this information and display it in the conventional way with dimension lines, etc, you will have to enter AutoCAD's "DIM" mode. This is a sub-program of AutoCAD which is used to produce all the dimension lines semi-automatically and interactively. All the types of dimensioning normally found on engineering and architectural drawings are catered for, and as with the rest of AutoCAD you have complete control over how it is drawn.

In this section you will add horizontal and vertical dimensions, a diameter and radius and add center markings for the flange bolt holes. Before actually drawing any dimensions change to the **Dimensions** layer. The dimension text will use the font of the current text style. In the Gland drawing the font, *isocp.shx* is used. Other fonts may give slightly different arrangements. If you successfully used the drawing template file Acadiso.dwt at the start of this chapter then your settings should be OK.

For ensured compatability set the STANDARD text style font isocp.shx. Pick **Format/Text Style...** and then from the font name pull down list, select **isocp.shx**. Then pick **Apply** followed by **Close**.

When you enter the DIM: environment only commands that help with dimensioning are allowed. All toggles and object snapping are available but many of the usual AutoCAD drawing and editing commands are not. If things go wrong, **Esc** key will always cancel the command and return you to the "Dim:" prompt. To execute commands other than dimensioning, you will have to exit from the "Dim:" prompt by pressing **Esc** key or typing **EXIT**. Picking commands from the pull-down menus will also exit the DIM: environment.

The actual dimensioning commands can be found in the **Dimension** pull-down menu (Figure 7.10). The most commonly used command, Linear, appears at the top of the menu. Linear allows horizontal and vertical dimensioning which will be used on the Gland. Aligned will be used later on a polygon object. Baseline and continue are useful for generating a line of dimensions or a set of running dimensions. Then there is a group of commands for drawing either a diameter or radius dimension and to dimension an anlge. A center mark can also be added. The **Ordinate** menu is useful for dimensioning points relative to some datum. Finally, the **Style** command controls how the dimensions are actually displayed.

To display the Dimension tool bar pick **View/Toolbars...** followed by **Dimension**. These buttons are the quickest way to do most dimensioning.

To dimension the flange of the gland pick **Dimension/Linear** or the **Linear Dimension** button. You are prompted for the "First extension line origin". Pick the leftmost point of the flange (point A on Figure 7.10). You could use the object snap Intersection for this. Pick the furthest right point (B on Figure 7.10) for the second extension line origin. After selecting the two points to be dimensioned you are asked where you want the actual dimension line to be drawn. Give any point whose y coordinate is 38 units.

Command: _dimlinear
First extension line origin or RETURN to select: **INT**
of **22,20**   (Leftmost point on flange)
Second extension line origin: **48,20**   (Other end of flange)

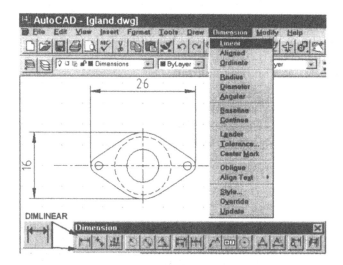

Figure 7.10 Dimension menu

Dimension line location (Mtext/Text/Angle...): **50,38**
Dimension text = 26

AutoCAD then calculates the dimension and zaps in the text, arrows and extension lines. This is an improvement on earlier versions of AutoCAD. If you want to have a little more control over the text that is written to the drawing try the following sequence.

The text is aligned with the dimension line and drawn above it. The "Mtext/Text/Angle" option in the "Dimension line location" prompt allows you to check the text before inputting a location and override the alignment by giving an angle for the text.

If your dimension is not correct, type **undo** to erase it and try again. If the text is too large or too small, carry on with the next vertical dimension and then read through the section on Dimension Style.

The "dimlinear" command produces the horizontal or vertical distance between any two points. It is DIM's version of using ORTHO. The horizontal distance is calculated from the X coordinates of the two extension line origins. The "vertical" operates in a similar fashion for vertical distances calculated from the Y coordinates.

Pick the **Linear dimension** button once more and give the object snap quadrant points at C and D as the extension line origins. The dimension line should be located at (8,22) or nearby. At the end of the sequence you are returned to the "Command:" prompt.

Command: _dimlinear
First extension line origin or RETURN to select: **quad**
of (Pick point C near 35,28)
Second extension line origin: **quad**
of (Pick point D near 35,12)
Dimension line location (Mtext/Text/Angle/...: **T**
Dimension text <16>: **<ENTER>**
Dimension line location (Mtext/Text/Angle/...: **8,22**

Dimension style

The template drawing, Acadiso.dwt, contains a number of different Dimension Styles. The one used above has been the default called "ISO-25". ISO stands for Internation Standards Organisation. It should have all the settings to produce the arrows and text shown in Figure 7.10. If you had problems reproducing the dimensions above then this section should fix matters. Your dimension style may have been different because of the settings in your workplace and AutoCAD setup.

AutoCAD controls the display of dimensions through a hideous number of variables and parameters. These can be accessed by typing:

Command: **DIM**
DIM: **STATUS**

This produces three screen-fulls of variables in a most un-userfriendly list. However, knowing that these variables exist gives an understanding of how the Dimension style dialog boxes work.

Even though the default settings of all the dimension variables will give reasonable looking results there will be times when you will want to make alterations. Once you have found your favorite combination it can be saved as a "Dimension Style". Indeed, using this feature you can gain access to all of the variables in a more user friendly fashion.

Hide the AutoCAD text window with all those dimension control variables and pick **Format/Dimension Style...** from the menu bar. We will create our own dimension style based on the ISO-25 default. Figure 7.11 shows the available styles. In the Name field, type **ISO-25-EXPRESS** and pick the **SAVE** button. This copies the ISO-25 style to the new name. We will now proceed to edit the new style.

The Dimension Styles dialog box has three sub-dialog boxes, namely, Geometry, Format and Annotation. Pick the **Geometry** button to bring up the dialog box shown in Figure 7.12. This defines the type of arrowhead to be used etc. The dialog is split into five sections. The "Dimension line" section allows

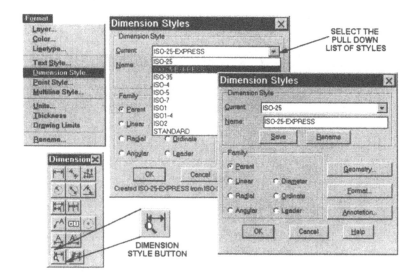

Figure 7.11 Dimension styles

the selection of the color of the line (which may be different from the color or the dimension text). You can also specify the spacing between lines and whether the line should be suppressed. The "Extension line" part of the dialog specifies the gap (origin offset) between the object being dimensioned and the start of the extension line and also the amount of extension beyond the end of arrow. Elsewhere you can specify the arrow size. Set the Center Mark size to **1.25**. This will used later.

The bottom section of the Geometry dialog box contains the most important parameter of all. The **Overall Scale** is probably the oneparameter you are most likely to need to change. All other variables are multiplied by this factor. If you have large limits on a drawing you may need to put in a large value for this scale factor. This alters the DIMSCALE variable. It increases or reduces the size of the components making up the dimension annotation, i.e. text height, arrow size etc. It does not affect the actual value of the dimension.

Pick **OK** to get back to the Styles dialog and then pick the **Format** button (Figure 7.13). Here you can control the way the text will be presented. Pick the **Vertical Justification** pull-down list and select **Centered**. This will change the text from being above the line to being in the line. The line will automatically have a break in it to fit the text. Pick **OK** to return to the Styles dialog once more.

Now pick the final button marked **Annotation** for the dialog box in Figure 7.14. Here you can change the text style or font to be used. You can

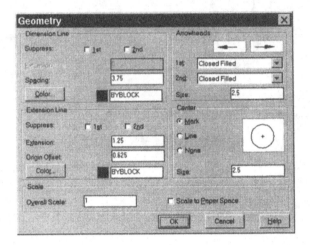

Figure 7.12 Dimension Geometry

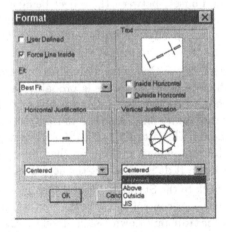

Figure 7.13 Dimension format

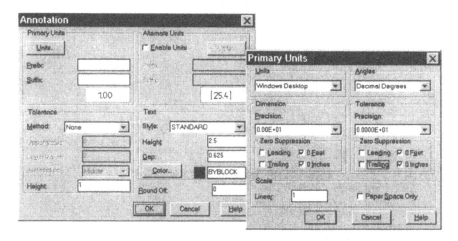

Figure 7.14 Dimension annotation

also enable alternate units (useful for converting between metric and feet and inches), use tolerance dimension and control the precision of the displayed dimension text.

Currently, in the GLAND drawing all the dimensions are shown with no values after the decimal point. Pick the **Units...** button to show the Primary Units dialog in Figure 7.14. Uncheck the suppression of trailing zeros for the dimension and alter the Dimension precision to **00.00E+01**. Note that the Linear Scale shown towards the bottom of the dialog is used to alter the value of the dimension text i.e. a value of 2 here would change the dimensions Figure 7.10 to 52 and 32. This factor should be used with great care. We need to keep it at unity.

Having done these changes pick **OK** to exit the Primary Units dialog and **OK** for the Annotation. Pick **SAVE** and then **OK** in the Dimension Styles dialog box to update the settings.

The two original dimensions will be unchanged. They were done using the style "ISO-25". The one we altered was "ISO-25-EXPRESS" and this is now the current dimension style.

subsection*Completing GLANDS dimensions

Having set the scene, let's see how it works out with our new style. To add the distance between the flange bolt holes and the radius of the smaller flange arc use **Dimension/Linear** once more followed by the **Continue** button. Pick the centers of the two bolt holes as the extension line origins and position the new dimension line below the "26", as shown in Figure 7.15.

Command: _dimlinear

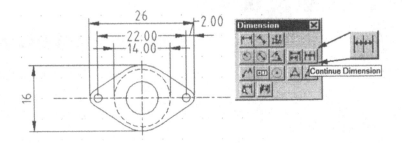

Figure 7.15 More horizontal dimensions

Dim: **horizontal**
First extension line origin or RETURN to select: **cen**
of (Pick bolt hole circle at 24,20)
Second extension line origin: **cen**
of (Pick right-hand hole at 46,20)
Dimension line location(Mtext/Text/Angle...): **45,35**
Dimension text = 22.00

Note the two zeros that now appear. To continue the dimension line to the right for the distance to the end of the flange pick **Dimension/Continue** from the menu bar or pick the **Continue Dimension** button.

Command: _dimcontinue
Specify a second extension line origin or (Undo/<Select>: **48,20**   (B)
Dimension text = 2.00
Specify a second extension line origin or (Undo/<Select>: **<ENTER>**

This should give the new horizontal dimension lines shown in Figure 7.15. The "2.00" is too long to fit between the extension lines so AutoCAD automatically places it outside. We will return to this dimension later to change its location slightly. The Continue Dimension can also be used for vertical dimensions.

You don't have to pick the two extension line origins if the object being dimensioned is made from just one entity. All you have to do is press <ENTER> and select the object. Now, rather than giving the extension line origins press <ENTER> to select the object and pick the circle.

Command: _dimlinear
First extension line origin or RETURN to select: **<ENTER>**
Select object to dimension: **30,25** (Point on dashed circle)
Dimension line location (Mtext/Text/Angle...): **30,32**

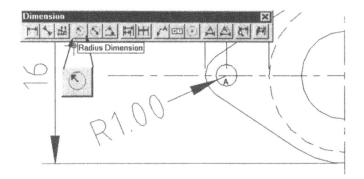

Figure 7.16 Radius

Dimension text = 14.0000 To demonstrate the use of the "radius" and

"diameter" commands, the right-hand bolthole and the central circle will now be dimensioned (Figure 7.16).

Pick **Dimension Radius** from the pull-down menu. Then pick anywhere on the small circle at the left-hand side of the gland. If snap and/or ortho are currently on toggle them off by double clicking the buttons on the status bar at the bottom of the screen. In the sequence given below a transparent zoom command is used to get a closer look. When the dimension command resumes, pick the circle and drag the dimension to a suitable orientation. As the text does not fit inside the circle it will be put outside.

Command: _dimradius
Select arc or circle: **'ZOOM** or pick View/Zoom/Window
>>all/Center/.../Window/<Realtime>: **W**
>>First corner: **7,10**
>>Other corner: **29,26**
Resuming DIMRADIUS command.
Select arc or circle: (Pick circle, A, near point 23.3,19.3)
Dimension text =1.00>
Dimension line location (Mtext/Text/Angle): (pick point near 12,14.5)

The apostrophe or quote before the zoom executes the transparent version of the command which means that it can be run in the middle of other commands including DIM commands. This command is located in the View pull-down menu and on the Standard toolbar.

Before including the diameter for the internal circle, use the zoom to return to the previous magnification. Pick **Diameter** from the **Dimension** menu or type "DIMDIAMETER" at the Dim: prompt. Then pick the inside

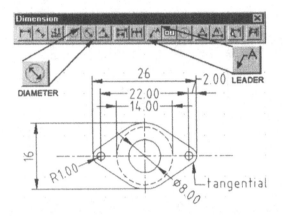

Figure 7.17 The dimensioned gland

circle near the point (38,17). Accept the text offered by AutoCAD and give a leader length of 6.

Command: **ZOOM**
All/.../Previous/.../<Realtime>: **P**
Command: _dimdiameter
Select arc or circle: **38,17**
Dimension text = 8.00
Dimension line location (Mtext/Text/Angle): a point near **52,14**

This time the default text causes the diameter mark ϕ to appear in front of the text. This is the standard symbol for indicating a diameter dimension.

To draw the center mark for the two bolt holes pick **Dimension/Center Mark** from the menubar and pick the circles near the points 24,19 and 46,19.

Command: _dimcenter
Select arc or circle: **24,19**
Command: _dimcenter
Select arc or circle: **46,19**

The size of this mark is dictated by the value of dimscale multiplied by dimcen which was set at the Geometry dialog box on page 188.

To conclude the gland drawing (Figure 7.17) put a leader line to indicate that the sloping line is tangential to the arcs. The **leader** command allows you to place pointers with text on the drawing. Once you have picked **Leader** from the **Dimension** menu you will be prompted for the leader start point.

This is where the point of the arrow will be drawn. Use the object snap **end** to locate the end of the arc and line near the point (47,18). The prompt changes to the familiar "To point:" request similar to the LINE command. Pick the point (51,16). You can make the leader line consist of as many line segments as necessary by picking points. When the line is completed press <ENTER> to exit the "To point:" prompt. Then type the desired text, which is the word "tangential".

> Command: _leader
> From point: **End** of **47,18** (or pick a point near to this)
> To point: **50,13**
> To point (Format/Annotation/Undo) <Annotation>: <ENTER>
> Annotation (or press Enter for options): **tangential**
> MText: <ENTER>

Even though you gave only two points on the leader line, leader line has a final horizontal line segment drawn automatically. The only real problem that can occur with drawing leader lines is if the distance between the start point and the first point is not long enough to draw the arrow. If this happens the command continues as normal but no arrow is drawn.

Editing Dimensions

Now that all the dimension lines and text have been added you can explore some of the editing facilities for associative dimensions. One of the undesirable features of the dimensions in Figure 7.17 is that the two places of decimals are not really necessary. The dimension text of "26" at the top and "16" on the left can be modified to read "26.00" and "16". Also the radial diameter text can be made horizontal and the "2.00" can be aligned with the 22.00 (Figure 7.18).

To set the radius text to be horizontal, pick **Format/Dimension Style** and pickthe **Format** button. Then put a tick in the box for "Outside horizontal". Now to update the diameter and the 2.00 with the new style setting. Pick the **Dimension Update** button or type:

> Command: **DIM**
> Dim: **UPDATE**
> Select objects: **pick the 2.00 text**
> 1 found
> Select objects: **pick the ϕ 8.00 text**
> Select objects: **pick the 26 text at the top dimension**
> Select objects: **pick the 16 text at the left dimension**
> Select objects: <ENTER>

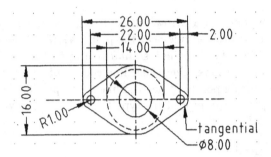

Figure 7.18 The updated dimensions

Only dimension entities will be modified with the **UPDATE** command. There are a number of other ways to edit dimensions. The Dimension Edit and Dimension Text Edit buttons are worth exploring (Figure 7.17). Text modification functions allow you to **Move Text** and **Rotate Text**. Both are self explanatory and useful when the drawing becomes congested. The **Hometext** option returns dimension text to the default position, based on the dimension style settings. Hometext also comes in handy after a dimension line has been STRETCHed.

The result of these alterations is shown in Figure 7.18. Note that you can move dimensions around very effectively by using their grip points. This is particularly useful in congested areas. There is a likelihood that the ϕ 8.00 might overwrite the leader text. If this is the case, simply select the diameter dimension and drag its grip a bit lower.

Standard dimension styles

Most drawing offices have a house style for dimensioning drawings. By setting all the parameters and creating a named dimension style you can quickly access the house style. In this drawing the International Standards Organisation settings were used (with a minor modification). In USA the ANSI standards are applicable while in Germany you woulds use DIN. The best way to access these standard dimension styles is by using the drawing templates when starting new drawings.

You can of course create your own dimension style from scratch or based on another standard and then save the drawing as a template file. To save a file as a template youjust pick File/Save As and then at the File type field you select "dwt".

It makes sense to create a number of consistent styles for the main type of drawing you normally produce. This takes a little thought and an analysis of your output. However, it saves hours in the long run. Typically, AutoCAD operators spend one third of their time dimensioning and annotating drawings. Setting good dimension styles can really spped up your drawing production.

Drawing a pentagon

To illustrate the remaining dimension commands and a couple of new drawing commands, you will now create a pentagon and find out the internal angle between two adjacent sides. To save time in setting up a new drawing environment just change to the POLYGON layer and freeze the others. This will allow you to use all the current dimension variable settings. Some of these will be altered temporarily to create tolerant dimensions. The quickest way to freeze all layers except the current one is to use the keyboard.

    Command: −LAYER
    ?/Make/Set/...:S
    New current layer <DIMENSIONS>: POLYGON
    ?/Make/.../Freeze/Thaw: F
    Layer name(s) to Freeze: *
    ?/Make/Set/...: <ENTER>

This method of freezing all the layers but the current one is much faster than using the Layer Control dialog box. AutoCAD will never freeze the current layer – that would be much too silly.

The **Polygon** can be found in the **Draw** pull-down menu. There are three choices for creating a polygon. It can be inscribed in a circle (the vertices touch the circle) of a given radius. It can circumscribe a circle (sides are tangential to the circle). Here we will specify the number of sides and the length on one edge.

When you select this command you are first prompted for the number of sides. Use 5 sides for a pentagon. You can then either specify a center of the polygon or you can give the position and length of one side. If you give the center, you will be asked whether you want to inscribe or circumscribe the circle

    Command: POLYGON
    Number of sides: 5
    Edge/<Center of polygon>: EDGE
    First endpoint of edge: 20,10   (A)
    Second endpoint of edge: @20,0   (B)

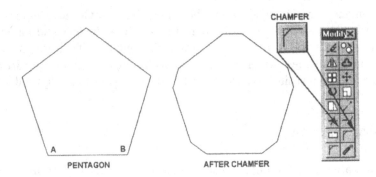

Figure 7.19 The chamfered pentagon

This actually draws a closed polyline, calculating the vertices from the geometrical properties of equilateral polygons. It can be edited in the same way as any other closed polyline.

The CHAMFER edit command can be used to cut off all the corners. This command is similar to the FILLET command but draws a straight line between the chamfer points. For this command you give the length by which each of a pair of lines is to be trimmed back. If a polyline is to be chamfered then you have the further option of trimming all the corners. For example, to chamfer the corners of the pentagon by trimming 3 units from each end of the line segments you would get the shape given in Figure 7.19. You first have to give the sizes of the chamfer and then the polyline to be edited. Pick **Construct/Chamfer.**

> Command: **CHAMFER**
> (TRIM mode) Current chamfer Dist1 = 10.0000, Dist2 = 10.0000
> Polyline/Distance/Angle/...<Select first line>: **DISTANCE**
> Enter first chamfer distance <10.0000>: **3**
> Enter second chamfer distance <3.0000>: **<ENTER>**
> Command: **<ENTER>**   (To re-execute the command)
> CHAMFER
> (TRIM mode) Current chamfer Dist1 = 10.0000, Dist2 = 10.0000
> Polyline/Distance/Angle/...<Select first line>: **POLYLINE**
> Select 2D polyline: **30,10**

Chamfer can also be used with unequal distances and be applied to individual pairs of lines. If a polyline to be chamfered contains an arc then the arc will be deleted and replaced with a straight line.

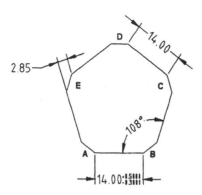

Figure 7.20 Final dimensions

Wrapping up dimensions

The "angular" option in the Dimension menu allows the dimensioning of angles
between lines. To draw the angle between the longer lines AB and BC, pick
Dimension/Angular. Then pick points on lines AB end BC. Indicate where
the dimension arc is to be located and accept the default dimension text and
text location.

> Command: _dimangular
> Select arc, circle,line: **30,10**   (Line AB)
> Second line: **43,20**   (Line BC)
> Dimension arc line location (Mtext/Text/Angle): **31,10**
> Dimension text = 108

The default text location causes the "108°" to be positioned in the middle of
the arc. The dimension should look like that in Figure 7.20.

With aligned dimensions the length is measured parallel to the line joining
the two extension line origins. To find the new length of the line between C
and D pick **Dimension/Aligned** from menu bar. Instead of picking the origin
points press <**ENTER**> and then pick the line CD at the point (38,35). Put
the dimension line at (42,39) and accept the default text.

> Command: _dimaligned
> First extension line origin or RETURN to select: <**ENTER**>
> Select line,arc, or circle: **38,35**
> Dimension line location (Text/Angle): **42,39**
> Dimension text = 14.00

The dimension line is aligned with the line segment and gives the correct length. The original length was 20 from which 3 was taken from each end.

To dimension the chamfer at the point E use the **rotated** dimensions at an angle of 198 degrees. This method is not supported by the menus or toolbars but still works. Type **DIM** and then **Rotated**, press <**ENTER**> and pick the short line at point E. Place the dimension line at (13,30) and accept the default text.

> Command: _dim
> Dim: **ROTATED**
> Dimension line angle <0>: **198**
> First extension line origin or RETURN to select: <**ENTER**>
> Select line,arc, or circle: **15.5,28.5**
> Dimension line location (Text/Angle): **13,33**
> Dimension text <2.85>: <**ENTER**>

Note that rotated dimensions are not hte same as aligned dimensions. Rotated dimensions were prominent in earlier releases of AutoCAD but are not well documented in Release 14. You can achieve the same effect with dimlinear if you rotate the snap angle.

Finally to produce dimensions with a tolerance level built in you need to change the Dimension style settings. Pick **Format/Dimension Style....** Click the ISO-25-EXPRESS style. Then pick the **Annotation** button. Now, pick **Tolerance Method** pull down list and select **Deviation** and give the upper value as **0.50** and the lower as **0.30** and the tolerance height at **0.5** (Figure 7.21). Then pick **OK**. Pick **OK** from the Dimension Style and use **dimlinear** to dimension the line AB.

Note that the new style for this last dimension all the dimensions will be applied to all the other dimensions if the update command is executed. It would therefore be a good policy to make a new style for the tolerance dimensions.

> Command: **dimlinear**
> First extension line origin or RETURN to select: <**ENTER**>
> Select object to dimension: **30,10**
> Dimension line location (Mtext/Text/...): **30,3**
> Dimension text = 14.00

This should give the dimensions as shown in Figure 7.20. If you override the default dimension text then the tolerance values will not be given.

To finish this exercise, pick **File/Save** from the menu bar followed by **File/Exit**.

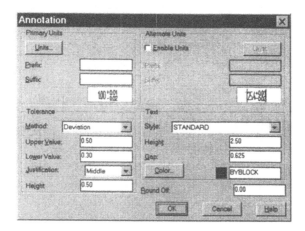

Figure 7.21 Tolerant dimensions

Summary

In this chapter you have encountered some advanced drawing and editing commands. Some of these, such as TRIM and CHAMFER allow you to dispense with having to draw preparatory construction lines. Others like POLYGON and DIVIDE draw multiple objects. By far the most important component covered in this exercise has been the dimensioning sub-system. AutoCAD changes when you are in Dim: and many new commands are made available while at the same time most of the drawing and editing commands are withdrawn.

Dimensions are calculated automatically from the current drawing units. It is important to choose suitable units and accuracy levels for sensible dimensioning. The dimension environment can be tailored to your specific needs by setting up an appropriate dimension style.

You should now be able to:

Draw lines tangential to two circles
Draw equilateral polygons
Trim and chamfer objects
Use transparent and realtime zooms
Add horizontal and vertical dimensions
Draw aligned, rotated, angular and tolerance dimensions
Set up a dimension style
Edit existing dimensions

Chapter 8 ADDING DEPTH TO YOUR DRAWINGS WITH 3D CAD

General

Three dimensional drawings come in three main flavors. Firstly, there are isometric projections that look like 3D but are merely 2D graphic constructions. These are useful for quick sketches and are commonly used in manual drafting. Isometric sketches also have their place in CAD as they are very efficient for conveying non-complex 3D information. The second flavor for creating 3D objects in CAD is the use of so called 2.5D shapes. These are 2D object that have an extruded thickness. The final flavor is the top of the range are full 3D surfaces and solids.

New users of 3D AutoCAD should find the tools and constuction techniques for 3D easy to pick up. Users upgrading from Release 12 Advanced Modeling Extension will have a bit of re-learning to do. The AME solid modeller was dropped in Release 13 to be replaced by the more advanced ACIS code. This means that there are new commands and many old 3D commands have been removed.

This chapter will take a brief look at all three techniques to produce some simple drawings. These include an isometric cooker, an office building and the great pyramid of Giza (Figure 8.1).

The opening up of the Z axis brings new and exciting aspects to AutoCAD use. Things can be constructed on the computer screen at full scale and depth. Once the object has been drawn it can be viewed from above (plan), from the front and side (elevation) and in either isometric or perspective projection. You can "walk" around the AutoCAD image and even through it. These facilities are particularly useful for disciplines where it is necessary to have a full appreciation and visualisation of the design.

As a cautionary note, one should not get carried away with the novelty and hype associated with 3D CAD. Architects and engineers have successfully managed to develop the most complex of projects over hundreds of years using simple 2D drawings. Thus, for a lot of design projects the 2D representation is adequate. Any changeover to 3D CAD must justify the extra effort required. You should also be aware of what AutoCAD 3D can and can't do.

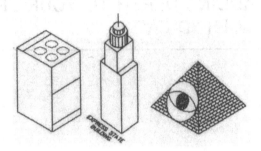

Figure 8.1 Pyramids and Towers

AutoCAD models surfaces and solid objects as wire frame skeletons. As such you can see through the objects that are drawn. You can tell AutoCAD to HIDE the lines at the back of the object to give the impression of solidity.

All the commands you have used up to now also work in 3D, although some have special versions when used in 3D (e.g. 3DPOLY replaces PLINE). There is a whole new vocabulary of terms relating to 3D geometry and a set of completely new functions. Let the work commence.

Isometric projection

Isometric projection is still the standard method of conveying three- dimensional engineering information on a two-dimensional sheet of paper. To produce anything other than simple shapes in isometric projection requires considerable expertise in drafting techniques. It is not my purpose to introduce such drawing construction methods but I do wish to display the special features within AutoCAD for isometric projections. To demonstrate these features and the basics of isometric projection we will create a drawing of the cooker that was used in Chapter 6.

Start up AutoCAD, pick **Start from Scratch** with **Metric** units. If AutoCAD is already running pick **File/New** to create a new drawing. Call it **COOKISO** with ACAD as the prototype. Set the LIMITS to (0,0) and (3250,2250) and the UNITS to Decimal. We will then set the GRID to 100 units and SNAP to 50 using the Drawing Aids dialog box.

Command: **LIMITS**
Reset Model space limits:
ON/OFF/<Lower left corner> <0.00,0.00>: <**ENTER**>
Upper right corner <420.00,297.00>: **3250,2250**

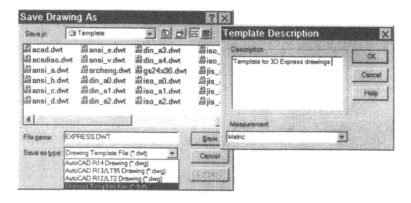

Figure 8.2 Saving a template file

Command: **ZOOM**
All/.../Window/<Realtime>: **A**

Setting units gives a lot of dialog which is truncated below. You can, of course use **Format/Units...** as described in earlier chapters.

Command: **UNITS**
Report formats:...
Enter choice, 1 to 5 < >: **2**
Number of digits to right of decimal point (0 to 8) <2>: **1**
System of angle measure:...
Enter choice, 1 to 5 < >: **1**
Number of fractional places for display of angles (0 to 8) <4>: **2**
Enter direction for angle 0.00 <0.00>: **<ENTER>**
Do you want angles measured clockwise? **<N> <ENTER>**

These settings will be useful for the other two drawings in this chapter. To keep them safe, SAVE the drawing as a template file with the filename **EXPRESS.DWT**. This will be used as a template for a later drawing. It contains the correct limits and unit settings.

Command: **SAVE**

Since this is the first save operation in the session the "Save Drawing As" dialog box appears. Pick **Drawing Template File (*.dwt)** from the pull down list of file types (Figure 8.2). Then give the name **EXPRESS.DWT** and pick the **Save** button. This brings a second dialog box for you to enter a Template Description. Type the text shown in Figure 8.2 and pick **OK**.

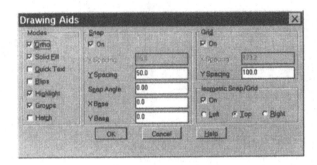

Figure 8.3 Setting Isometric Snap and Grid

The drawing title at the top of the AutoCAD window should now show "Express.dwt". You need to save the file one more time in order to change the name to COOKISO.DWG. Pick **File/Save As** from the menu bar. You may need to navigate back to the drawings folder and select the file type as "AutoCAD R14 Drawing (*.dwg)". Then give the drawing name as "COOKISO" and pick the **Save** button.

There is a subtle difference between typing SAVE and using the File menu. By typing the SAVE command, the current drawing name in the title bar, at the top of the window remains unchanged. If you pick File/Save As the title would change to the new name.

Now to set the Grid and Snap and to switch on the isometric projection pick **Tools/Drawing Aids...** from the menu bar. When the dialog box (Figure 8.3) appears, click the Snap **On** with a value of **50** for X and Y. Similarly, set the Grid to **On** with **100** for the X and Y values. Then pick the **On** box for **Isometric Snap/Grid** and pick the circle button beside **Top** as shown.

As soon as Isometric is switched on the X spacings for snap and grid change to 86.6 and 172.2 and become ghosted. The values of the X spacing in isometric are restriced to twice the Y value multiplied by the sine of 60°. This is because the "horizontal" isometric axes are 60° from the vertical or Y axis. Pick **OK** to accept the settings.

The three isometric projection planes are shown in Figure 8.4 but won't appear on your screen display. The X,Y and Z axes are at 150, 30 and 90 degrees from the horizontal. The orientation of the cursor cross hairs depends on which plane you want to work in. The effect of ORTHO also depends on the plane. The isoplane cube shown in Figure 8.4 defines the planes as LEFT, RIGHT and TOP. You can switch between the planes by pressing ^E or by using the Tools/Drawing aids dialog box. There is also a command, ISOPLANE, which does this. The orientation of the cursor cross hairs depends on the isoplane setting.

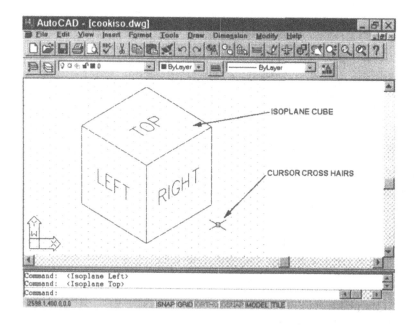

Figure 8.4 Isometric screen

Make sure that ORTHO is ON (Figure 8.3) and that coordinate display is set to polar mode. Toggling ^D twice should do this. Alternatively, you can set the value of the system variable, COORDS, to 2. Then switch to the right-hand plane to draw the front of the cooker shown in Figure 8.5(a). You will find it easier to drag the line points than to key them in. Watch the coordinate display for the correct lengths. The exact location of the first point is not too important. However, the relative positions of all other points are. If you make a mistake picking points use the u facility in the LINE command to undo that segment.

Command: **COORDS** or use ^D twice
New value for COORDS <1> **2**
Command: **ISOPLANE** or use ^E
Left/Top/Right/<Toggle>: **R**
Command: **LINE**
From point: pick a snap point near (1689,475)   (A)
To point: **@950<90** or drag along the "Iso-Z" axis   (B)
To point: **@500<30** or drag along "Iso-Y" axis   (C)
To point: **@950<270**   (D)
To point: **CLOSE**

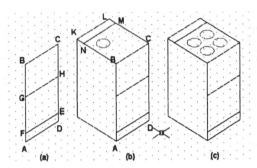

Figure 8.5 An isometric cooker

The next line, EF, is 100 units above the line DA. The point E is @0,100 from D. The line GH is a further 450 units above EF.

Command: **LINE**
From point: **@0,100**   (E)
To point: **@−500<30**   (F)
To point: **<ENTER>**
Command: **<ENTER>**
LINE From point: **@0,450**   (G)
To point: **@500<30**   (H)
To point: **<ENTER>**

Now switch to the left-hand plane to draw the side. You can use the ISO-PLANE again or use the toggle key ˆE. Pressing ˆE once changes to the left plane. You can cycle through all the planes quickly using ˆE. This toggle is also transparent so you can switch in the middle of another command.

Command: **ˆE** <Isoplane left> **LINE**
From point: **int** of pick point A using object snap.   (A)
To point: **@600<150**   (J)
To point: **@950<90**   (K)
To point: **int** of pick point B.   (B)
To point: **<ENTER>**

Use ˆE to toggle to the top plane to finish the cooker as shown in Figure 8.5(b and c).

Command: **ˆE** <Isoplane top> **LINE**
From point: **int** of pick point K.   (K)
To point: **@500<30**   (L)
To point: **int** of pick point C   (C)

```
To point: <ENTER>
Command: <ENTER>
LINE From point: @500<150 This is a point on CL 100 from L    (M)
To point: @500<210                                            (N)
To point: <ENTER>
```

To draw the heating elements you have to distort the circles. Luckily, the ELLIPSE command is just right for the job. The elements are at 200mm centers. The center of each circle is 150mm from the nearest edge. First, change the object color to red. You will then use the Isocircle option in the ELLIPSE.

```
Command: COLOR                      or pick the COLOR pull down list.
New object color <BYLAYER>: RED
Command: ELLIPSE or pick Draw/Ellipse/Axis, End
<Axis endpoint 1>/Center/Isocircle: Iso
Center of circle: pick the snap point near (1516,1675)   (3 snaps from N)
<Circle radius>/Diameter: 75              (The back left heating ring)
Command: COLOR
New object color <1 red>: BYLAYER               (Reset to normal)
```

AutoCAD uses the isoplane setting to calculate the correct amount of distortion and the orientation of the ellipse. This is a special feature of the ELLIPSE command, triggered when the SNAP style is isometric. Unfortunately, the ARRAY command does not support the isometric planes and so you have to use the simple COPY command.

```
Command: COPY
Select objects: LAST 1 found
Select objects: <ENTER>
<Base point or Displacement>/Multiple: M
Base point: 0,0                                 (Any point will do)
Second point of displacement: @200<30         (The back right ring)
Second point of displacement: @200<-30        (The front left ring)
Second point of displacement: F8 function key <Ortho off> @346.4¡0
Second point of displacement: <ENTER>
Command: QSAVE
```

The absolute X,Y coordinates that appear on the status line don't mean much when you are working in isometric projection. What is important is the relative position from the last point. Remember that the lines with <150 are parallel to the Iso-X axis, those at <30 are supposed to represent the Iso-Y axis and the vertical direction is Iso-Z. The cooker in Figure 8.5(c) is only a projection of the 3D information, it is not a 3D object.

The usefulness of AutoCAD's isometric projection will depend on the user's skill in that drafting technique. In general, you will have to use many construction lines to locate key points in the isometric view. Ortho, grid, snap and isoplane are very effective, while typing coordinates is not.

The Express State Building in 2.5 dimensions

It's time for a change of scene for all you out there, slaving over hot stoves. The next stop for the AutoCAD Express is the Big Apple where the skyline is about to be committed to the PC. In this example you will use the conventional 2D operations to draw a plan view. By also assigning a thickness and elevation in the vertical direction (Z axis) the shapes will have body as shown in the center of Figure 8.1.

Pick **File/New** to start a new drawing. Pick **Use a Template** and select **Express.dwt** from the list. If you didn't do the isometric cooker exercise, start a new drawing and follow the AutoCAD commands given above, down as far as the SAVE "Express.dwt".

To give depth to the drawing entities use the commands ELEVATION and THICKNESS. These allow you to set the altitude of the drawing plane and also the thickness or height of the entities. The main tower of the building shown in Figures 8.1 and 8.6 is a massive 500 units by 400 units and has a height, or thickness, of 1350 units. The height to the top of the mast is 2500 units.

A note for Release 12 and 13 users. In earlier versions of AutoCAD there was a dialog box for doing this. The command DDEMODES which activated that dialog box has been discontinued as most of the functionality is now in the object properties toolbar.

The default elevation is zero which means that the base of the tower is at ground level. Pick **Format/Thickness** from the menu to make the top of the first component at 1350. Then ZOOM in to the construction area and draw a rectangle for the main tower, ABCD in Figure 8.6. Set a SNAP value of 50 and GRID of twice this.

Command: **THICKNESS**
New value for THICKNESS <0.0> : **1350**
Command: **ZOOM**
All/.../Window.../<Realtime>: **W**
First corner: **700,700**
Other corner: **1800,1500**
Command: **SNAP**

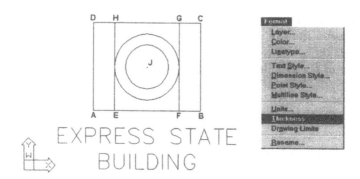

Figure 8.6 Plan view of skyscraper

Snap spacing or ON/OFF/Aspect/Rotate/Style < >: **50**
Command: **GRID**
Grid spacing(X) or ON/OFF/Snap/Aspect < >: **2X**

Now draw the base tower, ABCD,

Command: **LINE**
From point: **1000,1000**   (A)
To point: **@500,0**   (B)
To point: **@0,400**   (C)
To point: **@−500,0**   (D)
To point: **CLOSE**   (A)

The next part up, EFGH in Figure 8.6, is not quite as large but is 500 units
tall. Therefore the elevation and thickness must be changed. By setting the
elevation to 1350 the new entities will be drawn at the top of the main tower.
This can be done using the **ELEV** command. This combines the elevation and
the thickness in one command.

Command: **ELEV**
New current elevation <0.0>: **1350**
New current thickness <1350.0>: **500**
Command: **LINE**
From point: **1100,1000**   (E)
To point: **@300,0**   (F)
To point: **@0,400**   (G)
To point: **@−300,0**   (H)
To point: **C**

As the construction reaches skyward the elevation must be updated for the new elements. Climb to the top of the last object and draw a cylinder. A vertical cylinder is just a circle with a thickness. Note that the new elevation for the base of the cylinder is $1350 + 500 = 1850$.

> Command: **ELEV**
> New current elevation <1350.0>: **1850**
> New current thickness <500.0>: **250**
> Command: **CIRCLE**
> 3P/2P/TTR/<Center point>: **1250,1200**   (J)
> Diameter/<Radius>: **150**

And again. The new elevation is $1850 + 250 = 2100$.

> Command: **ELEV**
> New current elevation <1850.0>: **2100**
> New current thickness <250.0>: **100**
> Command: **CIRCLE**
> 3P/2P/TTR/<Center point>: **1250,1200**   (J again)
> Diameter/<Radius>: **100**

Now to draw the mast on top use a LINE with the full 3D cartesian coordinates. Above, you have input only the x and y values, allowing AutoCAD to deduce the z value from the current elevation. If you type the z value it overrides the elevation setting. The mast starts at the center of the top-most cylinder and extends 300 units into the air.

> Command: **LINE**
> From point: **1250,1200,2200**   (J)
> To point: **@0,0,300**   (300 straight up)

The last line command contains a deliberate error. Did you spot it? While the z value overrides the elevation, you cannot override the thickness setting. Thus the last line still had a thickness of 100 units. While it looks ok as a 2D dot, it is in fact 400 units high instead of the 300 required. There are two ways to fix this. Firstly, you could have simply drawn a dot or a point object. The second way is to modify the properties of the last line.

Pick the **Properties** button on the Object Properties toolbar (Figure 8.7). Then pick the line at (1250,1200,2200). This brings the Modify Line dialog box from which you can change the thickness from 100 to 0.

When you are finished working above ground level it is good practice to reset the elevation and thickness back to zero. This will help prevent user confusion if the drawing is made over a number of sessions. Once back to ground level, you can then add the title text.

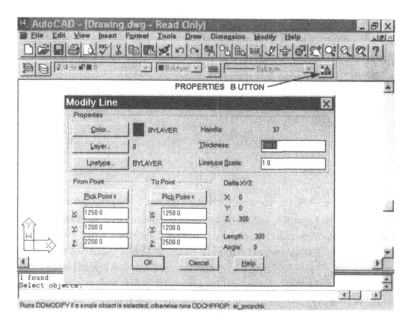

Figure 8.7 Modifying properties

Command: **ELEV**
New current elevation <2100.0>: **0**
New current thickness <100.0>: **0**
Command: **DTEXT**
Justify/Style/<Start point>: **C**   (Shortcut to centered text)
Center point: **1250,850**
Height <>: **70**
Rotation angle <0>: **<ENTER>**
Text: **EXPRESS STATE**
Text: **BUILDING**
Text: **<ENTER>**
Command: **SAVE** and give the drawing file name as **EXP-NY**.

Note, the centering of the text did not take place until the command input had finished. Your picture should now look like Figure 8.6.

Finally, to see the 3D effect of Figure 8.1 you will have to change the view point from which AutoCAD is looking. To get the solid effect you can remove the lines at the back. The commands VPOINT and HIDE do these jobs.

HAZARD WARNING! Always SAVE the drawing before a HIDE operation. On large drawings, the HIDE command can take a long time to calculate all the hidden lines to remove. Don't get impatient and start hitting the <ENTER> key. This only re-executes the last command, ie HIDE, and you will have even longer to wait. Use the Esc key to cancel key, if you want to interrupt. As there are only about 25 lines to be hidden in this drawing it shouldn't take more than a second.

Views and more views

The isometric projection view shown in Figures 8.8 is achieved by changing the view point so that we are looking at the tower from an angle. The actual viewing direction is parallel to the line joining this view point to the drawing's TARGET point. The default TARGET point is the origin. Note that the plan view point is $(0,0,1)$, ie looking down the Z axis from that point to the origin. A front elevation of the building could be generated with a view point of $(0,-1,0)$, a back elevation by $(0,1,0)$ and a side view by $(1,0,0)$.

To generate the isometric view you can use the 3D Viewpoint menu. Pick **View/3d Viewpoint** followed by **SE Isometric** (Figure 8.8). This sets the view point so that it appears we are viewing the building from a south easterly and elevated location. The actual value of the viewpoint is $(1,-1,1)$. The same effect can be achieved with the VPOINT command.

> Command: **VPOINT**
> Rotate/<View point> <0.0,0.0,1.0>: **1,−1,1**

Use the quick save command before executing the HIDE command.

> Command: **QSAVE**
> Command: **HIDE**
> Regenerating drawing.

This should give the required picture. It is a bit of a fraud, really. If you zoom in to the cylinders on the roof and redo the HIDE you should see that one of the lines at the top of the upper rectangular block is not correctly hidden. The reason for this is that the 2.5D lines produce an open ended rectangular box and not a solid block. The thick circles give solid cylinders though.

If you need to store particular view points you can use the DDVIEW command. This allows you to save the current display settings (VPOINT, ZOOM etc) for later retrieval. This is a help when trying to remember the VPOINT coordinates. Pick **View/Named Views...** from the menu bar. In the View Control dialog pick **New** and give the name as **Isometric-p**. Pick **Save View** for the current view (Figure 8.9).

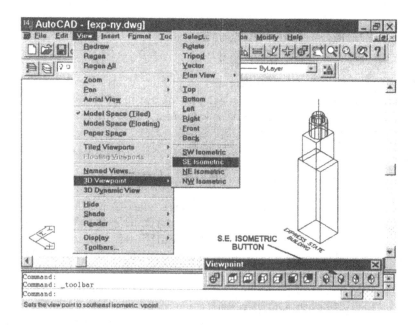

Figure 8.8 Viewpoint presets

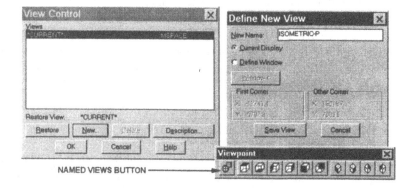

Figure 8.9 Saving Views

The keyboard equivalent of the dialog box is **View**.

Command: **VIEW**
?/Delete/Restore/Save/Window: **S**
View name to save: **ISOMETRIC-P**

The Viewpoint toolbar can be accessed by picking **View/Toolbars** from the menu bar followed by checking the **Viewpoint** box in the dialog. To show the front view type **VPOINT** and give a value of **0,−1,0**. Alternatively, pick the **Front View** button from the toolbar.

Command: **VPOINT**
Rotate/<View point> <1.0,−1.0,1.0>: **0,−1,0**

You can save this view if you wish. However, since AutoCAD has a menu item and button for the front view there is not much point. When you want to retrieve the saved view, pick **View/Named Views...** from the menu bar. This gives a dialog box with a list of all the saved views. Select **ISOMETRIC-P**, pick **Restore** followed by **OK**.

The other views that will be used in Figure 8.11 are the left side elevation and the plan. The left side view uses a viewpoint of -1,0,0 while the PLAN command or a viewpoint of 0,0,1 gives the latter. Do the **PLAN** command before proceding. Pick **View/3D Viewpoint** followed by **Plan View/World UCS**.

Command: **PLAN**
<Current UCS>/Ucs/World: **<ENTER>**

There is also a Plan button on the toolbar which only uses the World UCS (more of this later). Get to know the Viewpoint toolbar as it is really handy when navigating in 3D. The Named Views save not only the viewpoint but also the zoom magnification and center of the display. They are good for managing the display for plotting. With this objective, there will be more about Named Views in the next chapter.

Multiple views

There are two ways in which multiple views of an object can be shown. The first of these are the so called "tiled" viewports produced by the VPORTS command. This is essentially an old command whose functionality is available with many extras via the more modern floating viewports that result from the MVIEW command. As the VPORTS command may not be supported in future releases and as it offers nothing that MVIEW cannot do there is not much point in considering it. If you have old drawings that use VPORTS you

will have to consult *AutoCAD Express 2nd Edition* by yours truly or consult
the AutoCAD help file.

Model space and paper space

Drawings like the Express State Building can be considered as models of real
objects. There is a direct spatial correspondence between the computer model
and the real thing, that is, the model is drawn at full scale. For plotting
the model we have to produce a scaled image onto a piece of paper. Indeed,
engineering and architectural drawings usually show a number of views of the
object on one piece of paper.

Up to now we have created drawings solely in AutoCAD's "model space".
We will now set up a virtual page in "paper space" to view the four images
of the skyscraper. Paper space is a facility for viewing and plotting multiple
views so the full description of its capabilities is discussed in Chapter 9. The
floating viewports allow us to split the screen into a number of windows which
can be very useful in building 3D drawings. In this section only those aspects
needed to generate the windows will be described.

The first task in making these windows or metaviews is to move from
Model Space to Paper Space. Pick **View/Paper Space** from the menu bar.
This switches off the tilemode and moves from model space to paper space.
On the status line at the bottom of the screen you should notice that the word
"PAPER" now appears in place of "MODEL" and the TILE button appears
ghosted.

The first major effect of this is that the skyscraper disappears. Fear not,
it hasn't been deleted. Another thing you should notice is that the UCS icon in
the lower left corner of the screen changes from X and Y arrows to a triangle.
You are now in PAPER SPACE.

Paper space has its own LIMITS settings which should match the size
of paper used for plots. After setting the limits you will define a number of
viewports on this virtual page. These can be used to look at the skyscraper
from any viewpoint.

Command: **LIMITS**
Reset Paper space limits
ON/OFF/<Lower left corner> <0.0,0.0>: **<ENTER>**
Upper right corner < >: **420,297**
Command: **ZOOM**
All/.../Window/<Scale(X/XP)>: **A**

The floating viewports are made using the MVIEW command. The Floating
Viewports sub-menu shown in Figure 8.10 has a number of options. To make
the two tall, thin, viewports and the two squarer ones we will use "2 Viewports"

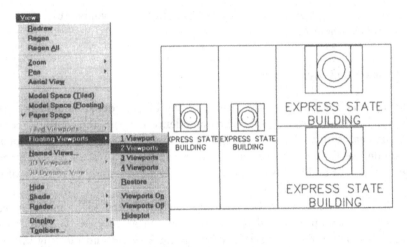

Figure 8.10 Making Floating Viewports

twice. Pick **View/Floating Viewports** followed by **2 Viewports** and follow the prompt sequence below.

    Command: _mview
    ON/OFF/Hideplot/Fit/2/3/4/Restore/<First point>: 2
    Horizontal/<Vertical>: **<ENTER>**

This means that the two viewports will be split by a vertical line. The following coordinates define the extremities of the two viewports combined. The area will be equally divided between them.

    Fit/<First point>: **10,10**   (A)
    Second point: **210,285**   (B)

When a viewport is first defined it will show the active model space view. In this case we left model space with a plan view showing. Now make the two squarer views.

    Command: _mview
    ON/OFF/Hideplot/Fit/2/3/4/Restore/<First point>: 2
    Horizontal/<Vertical>: **H**
    Fit/<First point>: **210,10**   (C)
    Second point: **410,285**   (D)

Each viewport is a window through which to view the model. All that remains now is to go back to model space and select the appropriate view for each

viewport. Note that PAPER SPACE is used to define the size and location of
the viewport but MODEL SPACE is used to define what is viewed. Double
pick the **PAPER** button on the status bar or pick **Model Space (Floating)**
from the **View** pull-down menu.

Command: MSPACE

The top right viewport should now appear with a heavier outline than the
others. This indicates that it is the "active" viewport. If you move the cursor
across the screen it will appear as cross hairs only in the active viewport.
Elsewhere, it will be a cursor arrow. If the top right viewport is not the active
one make it active by moving the cursor into it and press the mouse button.

To see the plan at a better magnification use the zoom command twice.
The first zoom fills the window while the second reduces the size to 90% of
the window size.

Command: **ZOOM**
All/.../Extents/Window/<Scale(X/XP)>: **E**
Command: **ZOOM**
All/.../Window/<Scale(X/XP)>: **0.9X**

Now move to the far left viewport, make it active and pick the **Left Side**
button from the Viewpoint toolbar.

Command: _vpoint Rotate/<View point> <0,0,1>: \ non *-1,0,0

Note that as you move the cursor from left to right in this viewport only the
Y and Z values of the coordinates in the status line changes. This is because
the X axis is perpendicular to that viewport.

Now move to the next viewport and press the **Front View** button on
the Viewpoint toolbar. When the view is selected only the X and Z values will
change in the status line. During the command AutoCAD echoes:

Command: _vpoint Rotate/<View point> <0,0,1>: \ non *0,-1,0

Finally, activate the lower right viewport and restore the ISOMETRIC-P view.
Pick the **Named Views** button or use the VIEW command as given below.
Use ZOOM and PAN as required to get the display right. It is easier to drag iso-
metrics than input exact coordinates. The "1000<45" below moves the tower
to the right since that view was defined with an angle of 315° (−45°).

Command: **VIEW**
?/Delete/Restore/Save/Window: **R**
View name to restore: **ISOMETRIC-P**
Command: **ZOOM**

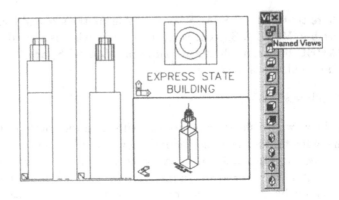

Figure 8.11 Four views of skyscraper

All/.../Extents/Window/<Scale(X/XP)>: **E**
Command: **PAN** drag tower to center of viewport if necessary.
Command: **QSAVE**

Viewports are very useful for 3D CAD by giving you an instant update of new entities in all the views. You can also switch between the viewports for the selection of entities or construction points in the middle of other commands. Their usefulness is not confined to 3D work and they can help speed up 2D drafting considerably. One viewport can be used to show a small-scale picture of the whole drawing, while other viewports can contain various details for working on.

Display commands such as GRID, ZOOM, VIEW, REDRAW and coordinate selection apply only to the active viewport. Two commands, REDRAWALL and REGENALL, cause all the viewports to be redisplayed. There is a limit to the number of viewports depending on your system. This is usually 48 and is controlled by the MAXACTVP variable. If you want to get rid of a viewport, you must go to paper space and use the ERASE command. There are many more features relating to MVIEW and PAPER SPACE. These are covered in some detail in Chapter 9.

The Pyramids of Giza in glorious 3D

Pack your bags and board the AutoCAD Express for your next destination, the ancient and three-dimensional land of Egypt. You have probably recognised from the previous section that there is a new level of complexity when trying to control points in 3D. In this section you will learn how to master this and construct a fully three-dimensional object, the Cheops pyramid.

The most difficult aspect of working in 3D is the optical illusion you encounter because the screen is only two-dimensional. To help with this problem you can set up a viewport for visualization and give it a suitable VPOINT. You will also use the coordinate filters, .x, .y, .z, .xy which allow you to pick the x value from one window and the y and z from others.

AutoCAD does provide an easy to use command for generating a pyramid. However, for the purposes of introducing many features of working in 3D we will adopt a round about route. Think of it as a kind of AutoCAD archeological dig looking for hidden treasures in 3D.

Pick **File/New** to create a new drawing to be called **EXP-GIZA** using **EXPRESS.DWT** as the template. Set SNAP to 50, GRID to 100 and draw the plan for the pyramid shown in Figure 8.12. After that, use MVIEW to set up three viewports on the screen to watch the 3D take off.

Command: **SNAP**
Snap spacing or ON/OFF/Aspect/Rotate/Style < 10.0 >: **50**
Command: **GRID**
Grid spacing (X) or ON/OFF/Snap/Aspect <10.0>: **100**
Command: **LINE**
From point: **1000,1000**   (A)
To point: **@600,0**   (B)
To point: **@0,600**   (C)
To point: **@−600,0**   (D)
To point: **C**   (A)
Command: **TILEMODE** or double click the TILE button on the status bar
New value for TILEMODE <1>: **0**

Using the TILE button or TILEMODE is the same as **View/Paper Space**. Now set the limits for paper space to (420,297) as before and make the two left-hand viewports shown in Figure 8.12.

Command: **LIMITS**
Reset Paper space limits
ON/OFF/<Lower left corner> <0,0,0>: **<ENTER>**
Upper right corner < >: **420,297**
Command: **ZOOM**
All/.../Window/<Realtime>: **A**
Command: **MVIEW** or pick View/Floating Viewports/3 Viewports
ON/OFF/Hideplot/Fit/2/3/4/Restore/<First point>: **3**
Horizontal/Vertical/Above/Below/Left/<Right>: **R**
Fit/<First point>: **10,10**

Second point: **410,285**

This makes the larger viewport appear on the right-hand half of the screen. This will also be the new active viewport when you go to MODEL SPACE. Once in model space, use ZOOM W to make better use of the display.

Command: **MSPACE** or double pick the PAPER button on the Status bar
Command: **ZOOM**
All...//Window/<Realtime>: **W**
First corner: **900,900**
Other corner: **1700,1700**

Make the lower left viewport active by moving the cursor into it and pressing the pick button. Then change the VPOINT to give a frontal view. Move the the upper left and set up an isometric type of view.

Pick lower left viewport.
Command: **VPOINT** or pick the Front View button
Rotate/<View point> <0,0,1>: **0,−1,0**
Regenerating drawing. Grid too dense to display.
Command: **ZOOM**
All...//Window/<Realtime>: **0.7X**
Command: **PAN**
Drag the horizontal line until it appears near the bottom of viewport.
The press **Esc** key
Pick upper left viewport.
Command: **VPOINT** or pick SE Isometric View button
Rotate/<View point> <0,0,1>: **1,−1,1**
Command: **ZOOM**
All...//Window/<Realtime>: **0.7X**

With these views you should be able to see if the lines to the apex of the pyramid are being drawn correctly as shown in Figure 8.12. Depending on you monitor size you might need to adjust the zoom magnifications. There are four sloping lines to be drawn from the corners, A,B,C and D, to the apex, E which is 475 units above the center point. To explain the facilities a number of coordinate definition methods are used.

Command: **LINE**
From point: **1000,1000**   (A)
To point: **@300,300,475**   (E)

If the Z ordinate is not specified it is taken as the current elevation setting (i.e. zero). The point E is 300 units along the X axis, 300 along the Y axis and

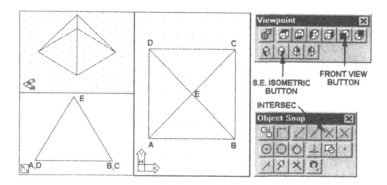

Figure 8.12 Cheops' pyramid

475 units up the Z axis from the point A. As the second point is input the line should appear in all three viewports. For the next point use object snap intersection to locate B in the right-hand viewport.

> To point: **intersec**
> of Pick the right-hand viewport and then pick the point B.
> To point: **<ENTER>**

The object snap is also able to pick up the full **XYZ** coordinates as demonstrated by the next sequence. You can also move between viewports during a command.

> Command: **<ENTER>**
> LINE From point: **intersec**
> of Pick point C
> To point: **intersec**
> of Pick point E

If you are not sure of the elevation of a particular point on the plan view you can use the .XY filter to use just those coordinates and type in Z separately.

> To point: **.xy**
> of **intersec**
> of Pick the point D
> (Need Z): **0**
> To point: **<ENTER>**

The complete pyramid should now appear similar to Figure 8.12 but without the letters A to E. Note that the bottom left viewport shows points A and D at the same location. This is correct as D is directly behind A. If you used this

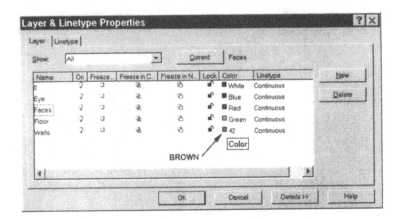

Figure 8.13 Layers of time

viewport for picking the point A the results might be uncertain because of the ambiguity. You may need to adjust the zoom in the left viewports.

Making the faces solid

The pyramid shown above is but the first step in construction. The 3D lines form only the frame on which we can hang the fabric. To make the slopes solid we will create 3DFACEs on each of the triangles. Making the 3D faces is like stretching fabric over the wire frame. Faces are opaque when the HIDE command is executed. At present, if you move to the upper left viewport and try HIDE, all the lines will still be visible.

Create the new layers shown in Figure 8.13. Assign different colors so that the faces are distinguishable from the original lines. The Walls layer has a brownish color (number 42 on my computer) selected from the Select Color dialog. Pick the color swatch in the Layer dialog to get this. Make sure that the current layer is set to **FLOOR** and that the right-hand viewport is active before starting on the faces. Then make a 3D face for the bottom of the pyramid and a face for each of the other four sides. When we get on to hatch the pyramid it will be useful to have the floor on a different layer to the other faces.

Command: **LAYER**

The 3DFACE command is found on the Surfaces toolbar or by picking **Draw/Surfaces/3D Face** from the menu bar (Figure 8.14). We will return

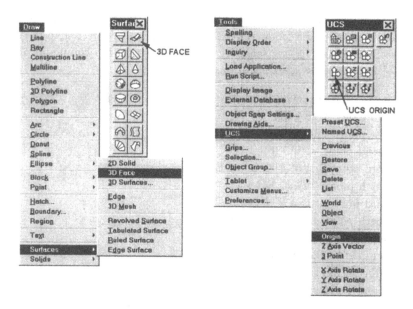

Figure 8.14 3D and UCS menus

to the other items on the 3D menu later. The first face is to be the floor of the pyramid.

Command: 3DFACE
First point: **1000,1000**   (A)
Second point: **int** of pick B
Third point: **int** of pick C

Let's use the filters for the last point, for fun. Point D has the same X value as A and the same Y and Z as C.

Fourth point: **.x**
of **int** of Pick point A
(need YZ): **int** of Pick point C
Third point: **<ENTER>**

You will be prompted for more third and fourth points to add more faces onto the last edge. Note that only the red edges of the face are shown. Faces are never filled but they are opaque when using **HIDE**. The four points defining the face should be on the same plane if possible. It is not an error to use non-coplanar points but it is sloppy **3D CAD**. The picture will look the same as Figure 8.12 but the square should now be in green.

Now make the sloping faces ABE and CDE in one 3DFACE "bow-tie" operation. This will be followed by BCE and ADE. Use object snap to get the intersection points. In this sequence, pay attention to the command prompt line so that you press ENTER at the correct times.

Use the Layer control to switch to the FACES layer.
Command: **3DFACE**
First point: **1000,1000**   (A)
Second point: **@600,0**   (B)
Third point: **int** of Pick point E
Fourth point: **<ENTER>**
Third point: **int** of Pick point C
Fourth point: **int** of Pick point D
Third point: **<ENTER>**
Command: **<ENTER>**
3DFACE First point: **1000,1000**   (A)
Second point: **@0,600**   (D)
Third point: **intersec** of Pick point E at the apex
Fourth point: **<ENTER>**
Third point: **intersec** of Pick point C
Fourth point: **intersec** of Pick point B
Third point: **<ENTER>**
Command: **SAVE** and give the name **EXP-GIZA**

Faces always have four points defining them. To make a triangle two of the points must have the same location (eg first and second or third and fourth). Note that the third and fourth points of the previous face are used as the first two points in the next. This allows you to build up complex surfaces from a series or triangle and quadrilateral faces.

Your pyramid should still look much the same as before, but this time in red. To see the difference between the 3DFACE representation and the LINEs, move to the upper left viewport and issue the HIDE command. Then make the layer containing the lines the current layer and freeze layer, FACES. Try HIDE once more and you should still be able to see all the lines. Thaw and Set the FACES layer again for the next part of the exercise.

Define your own coordinate system

The most important features in 3D AutoCAD is the availability of user definable coordinate systems. This means that you can reset the position of the origin and also the orientation of the X, Y and Z axes. The default coordinates system that has been used up to now is the World Coordinate System (WCS).

The WCS specifies the drawing origin and the directions of X, Y and Z axes. Other coordinate systems are defined relative to this.

One possible point of difficulty can be deciding on which direction the positive Z axis points towards. AutoCAD uses the right hand rule to define all coordinate systems. Place your right hand near the computer screen with your palm facing you and extend the thumb to the right, forefinger up and middle finger towards you. These fingers show the positive directions of the X, Y and Z axes respectively. If you keep your fingers in that postiion and rotate your hand you will see how the axes of the new coordinate system relate to each other.

In this section you will draw an inscribed circle on the ABE slope. To try this in the WCS would be fruitless because AutoCAD circles are always drawn in the XY plane. You have to define a User Coordinate System (UCS) parallel to the slope. In fact you have to make a new UCS for every new plane you want to draw circles or other 2D entities such as HATCH on.

The UCS command appears on the Tools pull-down menu. There is also a UCS toolbar. Pick **Tools/UCS** to see the menu shown in Figure 8.14. Pick **Origin** to reset the origin to the point A on the pyramid.

Command: **UCS**
Origin/Zaxis/.../?/<World>: **O**
Origin point <0,0,0>: **1000,1000,0**   (A)

If you pick **View/Display** followed by **UCS Icon/Origin** the little UCS icon should move to A in the active viewport. This uses the UCSICON command. If you look closely, you will see that the "W" has disappeared from the icon. A small "+" indicates that it is at the UCS origin. This shifting of the origin can be very useful even in 2D drawings. Note that all coordinate values are now relative to this new origin.

Now define the four slopes as new coordinate systems. The options for defining the plane are quite varied. You can align the UCS with an object such as a 3DFACE, or you can specify three points in the plane. You can also select the XY plane or specify a new Z axis direction. The UCS can also be set to a particular VIEW or rotated about any of the XYZ axes. Different UCS definitions can be named and saved and restored like VIEWs.

Command: **UCS**
Origin/.../OBject/.../Save/Del/?/<World>: **OB**
Select object to align UCS: Pick the face ABE along edge AB.

The UCS icon should now take up its new orientation. The position of the origin is dependent on which point of the 3D face was originally drawn first. This method cannot be used with objects that contain non-coplanar points.

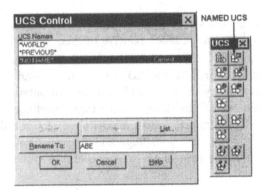

Figure 8.15 UCS Control

Now save this UCS. Pick **Tools/UCS** and **Named UCS....** The UCS control dialog box then pops up (Figure 8.15). The UCS that has just been defined is the current one. It appears on the list as "*NO NAME*". To give it the name "ABE" pick the line, "*NO NAME   Current" and type ABS in the box near the bottom as shown in Figure 8.15. Then pick the **Rename To:** button followed by **OK**.

Define the BCE plane by picking each of the three points with object snap intersection and save it. Pick **Tools/UCS/3 Point**.

Command: **UCS**
Origin/Zaxis/3point/...<World>:**3**
Origin point <0,0,0>: **int** of Pick point B.
Point on positive portion of X axis <601,0,0>: **int** of Pick C.
Point on positive-Y portion of the UCS XY plane <599,0,0>:
   int of Pick E.
Command: **UCS**
Origin/.../Save/Del/?/<World>:**S**
?/Desired UCS name: **BCE**

To demonstrate this method further restore the WCS and define a new UCS for the side CDE. Remember the right hand rule for positive axis directions. Pick **Tools/UCS/Preset UCS**. Then pick the World coordinate system icon, **top left**, from the UCS Orientation pop-up (Figure 8.16). This pop-up also gives quick access to the previously used UCS and allows you to set the current view as a UCS. The other presets are useful for swopping from the front to the back of an object or from the front to the side. These presets, however, are better for rectangular buildings than for the pyramid.

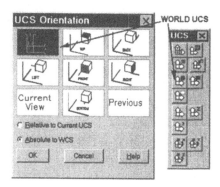

Figure 8.16 UCS Orientation

You can also set the world coordinate system with the UCS command or picking the World UCS button on the UCS toolbar.

Command: **UCS**
Origin/.../Save/Del/?/<World>:**W**
Command: **UCS**
Origin/ZAxis/3point/...<World>:**3**
Origin point <0,0,0>: **1600,1600,0**   (C)
Point on positive portion of X axis <1601,1600,0>: **int** of pick D
Point on positive-Y portion of the UCS X-Y plane <1600,1599,0>:
    1300,1300,475   (E)
Command: **UCS**
Origin/.../Save/Del/?/<World>:**S**
?/Desired UCS name: **CDE**

Here, the X axis points from C to D with Y pointing up the plane. Finally, use the Object option to make the fourth UCS for side DAE.

Command: **UCS**
Origin/.../OBject/...<World>:**OB**
Select object to align UCS: Pick the face DAE along edge AD.

The UCS icon moves to point A since that was the first point used to originally draw the 3D face. The X axis is positive along the line AD as that was the original input order of the points. The positive Y axis is up the face. From the right hand rule this means that the Z axis is positive into the pyramid. All the other faces have UCS's with Z positive out from the pyramid. To make this last UCS consistent with the others, you should move the origin to D and make D to A the positive direction for the X axis. This latter task can be accomplished

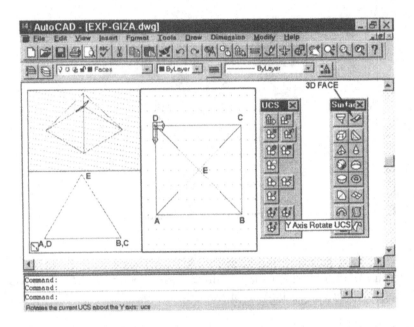

Figure 8.17 Rotating UCS about Y-axis

by rotating the UCS about the Y axis. Pick **Tools/UCS/Origin** to shift the origin 600 units along the X- axis to D.

```
Command: _ucs
Origin/.../X/Y/Z/...<World>:O
Origin point <0,0,0>: 600,0                    Point D in current UCS.
```

Now pick **Tools/UCS/Y Axis Rotate** to spin the X and Z axes 180° about the Y-axis to give the situation shown in Figure 8.17. This effectively flips the UCS over. You should now see the icon at D with the X arrow pointing towards A and Y pointing up the slope. Then save DAE.

```
Command: _ucs
Origin/.../X/Y/Z/...<World>: _y
Rotation angle about Y axis <0.0>: 180
Command: UCS
Origin/.../Save/Del/?/<World>:S
?/Desired UCS name: DAE
```

You have now created all the coordinate systems needed for drawing shapes on each of the pyramid faces. You can restore any of the named UCS's, or

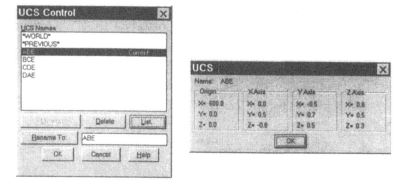

Figure 8.18 Setting UCS to ABE

delete ones that are no longer required from the UCS Control dialog box
(Tools/UCS/Named UCS from the menu bar.

The all seeing eye

You can now use these four coordinate systems to add bricks to the pyramid
walls and to draw the "all seeing eye". Use the right hand viewport for the
following constructions. If it is not already the active viewport then move the
cursor into the right hand viewport and press the mouse button.

Pick **Tools/UCS/** and **Named UCS...** to get the UCS Control dialog
box shown in Figure 8.18. Then pick **ABE** from the list. If you pick the **List**
button you get the details of ABE relative to the current UCS. Pick **OK** to
get back to the control dialog and with ABE highlighted pick the **Current**
button and **OK**. To set up a plan view before drawing the eye, pick **View/3D
Viewpoint** followed by **Plan View/Current UCS** as shown in Figure 8.19.

Moving to a plan view causes the view point to change so that you are now
looking perpendicularly down on the face ABE. AutoCAD executes a ZOOM
Extents automatically when a PLAN has been selected. This means that the
edge lines will appear at or near the very edge of the viewport. To get a better
picture use **ZOOM 0.9X** to reduce the size.

Command: **ZOOM**
All/.../Window/<Scale (X/XP)>: **0.9X**

Then switch to the **EYE** layer that was created earlier. Use the layer pull down
list for this. To inscribe a circle in the ABE triangle pick **Draw/Circle/Tan,Tan,
Tan.**
This automatically uses object snap "TANgent" for each point.

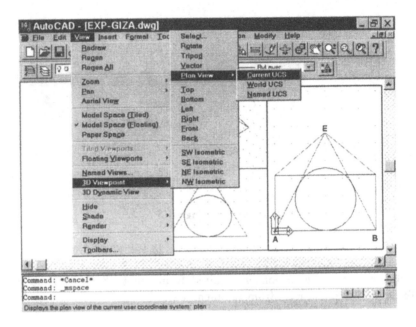

Figure 8.19 Inscribed circle

Command: _circle 3P/2P/TTR/<Center point>: 3p
First point: _tan to Pick the line **AB**.
Second point: _tan to Pick the line **AE**.
Third point: _tan to Pick the line **BE**.

The circle should fit nicely in the triangle. You can now use two arcs to draw
the eye shape. Use the quadrant points of the circle with object snap to locate
the end and center points. Then mirror the arc about the line FG (Figure 8.20).
Pick **Draw/Arc** followed by **Start Center End**.

Command: _arc
Center/<Start point>: **QUAD** of Pick circle near point F.
Center/End/<Second point>: _c
Center: **QUAD** of Pick bottom of circle at (300,0)
Angle/Length of chord/<End point>: **QUAD** of
Pick circle on its left, near point G.
Command: **MIRROR**
Select objects: **LAST** 1 found
Select objects: **<ENTER>**
First point of mirror line: **INT** of Pick point F
Second point: **@1,0**   (A horizontal line)

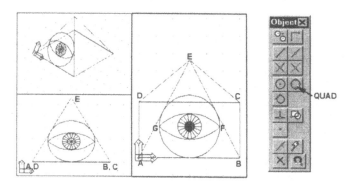

Figure 8.20 The all seeing eye

Delete old objects? <N>: <ENTER>

The pupil and iris will complete the all seeing eye. A donut with no hole makes a good pupil while a circular array gives the iris. Another circle encloses both. Use the **Draw** pull down menu or type the commands.

Command: **CIRCLE**
3P/2P/TTR/<Center point>: **CEN** of Pick the inscribed circle.
Diameter/<Radius>: **70**
Command: **DONUT**   (Can also be spelt "DOUGHNUT".)
Inside diameter <0.5>: **0**
Outside diameter <1>: **60**
Center of doughnut: **@**   (Last point i.e. the center of the circle)
Center of doughnut: <ENTER>

You are prompted to place more donuts. Pressing <ENTER> exits the command. If the donut looks facetted rather than smooth, try to zoom in and REGEN the view. Then do a zoom previous to get back. The donut will appear solid only in the plan view.

Now draw the line from the donut to the small circle and ARRAY it. Array is on the **Modify** menu. While this is a 3D drawing, this array is only a 2D one in the plane of the current UCS. There is a separate 3D Array command for working out of the plane.

Command: **LINE**
From point: **@30,0**   (30 to the left of the donut's center.)
To point: **@40,0**
To point: <ENTER>
Command: **ARRAY** or pick **Modify/Array**

Select objects: **LAST** 1 found
Select objects: **<ENTER>**
Rectangular or Polar array (R/P): **P**
Center point of array: **CEN** of pick either of the circles.
Number of items: **18**
Angle to fill (+ =ccw, − =cw) <360>: **<ENTER>**
Rotate objects as they are copied? <Y>: **<ENTER>**

The all seeing eye of Cheops' pyramid should now look like that in Figure 8.20. Now is a good time to SAVE the drawing as the next task is to HATCH the walls with a brick pattern. There is a potential problem in selecting the faces to hatch. All the red lines on the drawing belong to two faces. If you select a face by picking its edge you may get the adjacent face instead. To avoid this you can freeze the FLOOR layer and pick each of the wall faces using the bottom line.

Command: **QSAVE**
Command: **−LAYER** or pick the Layer Control pull down list
?/Make/Set/...: **S**
New current layer <EYE>: **WALLS**
?/Make/Set/...: **FREEZE**
Layer to Freeze: **FLOOR**
?/Make/Set/...: **<ENTER>**

The hatching operation firstly involves selecting the pattern, the scale and the hatching style. Pick **Draw/Hatch...** to get the Boundary Hatch pop-up screen (Figure 8.21). Pick the **Pattern...** button from the dialog box. Scroll down the list of patterns until you locate the one called **BRICK**. Then pick the rectangle of bricks just above the word "brick". The dialog reverts to the Boundary Hatch box. Having selected the pattern, you must now specify the scale as **4.0**. Check that the hatch style is normal by picking **Advanced**. When everything is satisfactory pick **OK** to return to Figure 8.21.

Now click the **Select Objects** button and pick the face, ABE, along the bottom line and the larger circle.

Select objects: pick the larger circle.
1 found
Select objects: **140,0** or pick the face ABE along the line AB.
1 found
Select objects: **<ENTER>**

AutoCAD returns to Figure 8.21 for further instructions. Before executing the hatch, pick **Preview Hatch** to see what it will look like. If it looks like

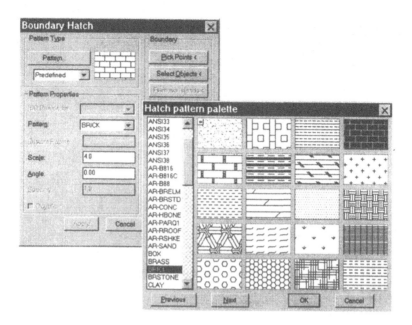

Figure 8.21 Selecting brick pattern

the lower left view in Figure 8.22 then press **<ENTER>** and pick **Apply**. Otherwise, go through the above procedure to fix any errors.

The area outside the circle should now be hatched in all three viewports. The hatch only becomes part of the drawing when the Apply button has been picked. If the wrong area has been bricked in use "ERASE, last" and try again.

To put the bricks on the other three walls you will have to restore each UCS and execute the HATCH command. Thus, you now need to use the UCS command or Named UCS to restore the BCE coordinate system.

Command: **UCS**
Origin/.../Restore/.../<World>: **R**
?/Name of UCS to restore: **BCE**

Now execute the brickwork by typing the HATCH command.

Command: **HATCH**
Pattern (? or name/U,style) <BRICK>: **<ENTER>**
Scale for pattern <4.0000>: **<ENTER>**
Angle for pattern <0.00>: **<ENTER>**
Select objects: **140,0** or pick the face BCE along the line BC.
1 found

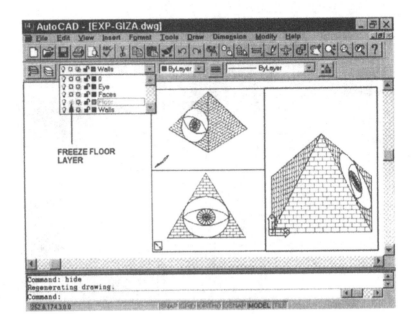

Figure 8.22 The brick-built pyramid

Select objects: **<ENTER>**

The bricks should now appear on the right hand wall. Note that even though the object is being viewed from the front, the hatch is always applied in the plan view and current elevation of the *CURRENT UCS*. This can be a bit disconcerting and often leads to errors. Therefore is is recommended that you execute a PLAN command after changing the UCS.

This is easier done than said, using the system variable, USCFOLLOW. Giving it a value of 1 causes AutoCAD to go automatically to the plan view whenever a UCS is restored or newly defined. Setting UCSFOLLOW to 1 enables this feature, 0 disables it. You will have to type it at the command line.

Command: **UCSFOLLOW**

New value for UCSFOLLOW <0>: **1**

Now do the remaining two sides. Note how the display goes straight to the plan view. Type the UCS command or pick **Settings UCS** and **Named UCS**.... Then restore CDE by picking it from the list and clicking the **Current** button followed by **OK**. The next sequence is the equivalent set of keystrokes. Do a

ZOOM to shrink the image. This will make it easier to pick the bottom line of the face.

Command: **UCS**
Origin/.../Restore/Save/Del/?/<World>:**R**
?/Name of UCS to restore: **CDE**
Command: **ZOOM** if necessary
All/.../Window/<Realtime>: **0.9X**

This time apply the hatch using the Boundary Hatch pop-up. Pick **Draw/ Hatch**.... All the settings from the previous hatch are still valid, so you can select the object, preview and then apply the hatch.

Command: _bhatch
Pick the **Select Objects** button.
Select objects: **140,0** or pick the face CDE along the line DE.
1 found
Select objects: <**ENTER**>
Pick **Preview Hatch** followed by **Apply**
Command: **UCS**
Origin/.../Restore/Save/Del/?/<World>:**R**
?/Name of UCS to restore: **DAE**
Command: **ZOOM**
All/.../Window/<Realtime>: **0.9X**
Command: **BHATCH**
Pick **Select objects** button
Select objects: **140,0** or pick the face DAE along the line DA.
1 found
Select objects: <**ENTER**>
Pick **Apply**

If you execute a HIDE in each viewport, your picture should look like Figure 8.22. Always save before using HIDE as it can take some time. Repeat the HIDE in each viewport.

Pick the upper left viewport.
Command: **QSAVE**
Command: **HIDE**
Regenerating drawing.

A dynamic view point on visualisation

Another of the goodies in AutoCAD is the Dynamic View command, DVIEW.
This is more versatile than the VPOINT command as it allows you to see the
object as you move and twist it in full 3D and in realtime. DVIEW effectively
combines ZOOM, VPOINT, and a perspective view option with a powerful user
interface. The emphasis is on **3D** visualisation and much of the terminology
comes from photography. To use DVIEW you have to imagine yourself looking
through a camera lens at a target point.

To see the pyramid in all its glory let's use the full screen and revert to the
World Coordinate System and TILEMODE, on. It is advisable to do a zoom
such that the object to be viewed appears near to the center of the screen at a
low magnification. If the object is near the edge or fills the screen the dynamic
view may cause it to go off screen completely.

Command: **TILEMODE** or double click the TILE button
New value for TILEMODE <0>: 1
Command: **UCS**
Origin/.../Restore/Save/Del/?/<World>:**W**

Back in tilemode you should see the pyramid in plan. This was the view when
tilemode was originally turned off. If your pyramid is not in plan then execute
the PLAN command. It is also necessary to thaw the FLOOR layer. Pick the
Layer pull down list and click on the snowflake by the Floor layer (Figure 8.22).
You should already see tha plan view. If not execute the plan command.

Command: **PLAN**
<Current UCS>/Ucs/World: **W**
Command: **ZOOM**
All/.../Window/<Scale (X/XP)>: **A**

Now pick **View/3D Dynamic View** from menu bar. You are then prompted
to select the objects to be viewed. Here you will select everything. Once the
selection has been completed you can begin to explore the many Dview options
for manipulating the display.

Command: _dview
Select objects: **All**
44 found.
4 were not in current space
Select objects: **<ENTER>**

The viewports exist in paper space and so are not in the current model space.
Only the selected objects will be shown in the dynamic previews. When the

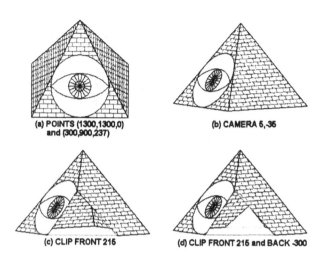

Figure 8.23 Dviews of Cheops with HIDE

final view has been chosen, all the drawing will be displayed. You could, for example remove the hatching from the selection set, which would speed things up slightly.

The command prompt changes to give all the display options. The display will be calculated relative to a given camera position and the target position. The target is the point where the camera is focussed on and will always end up in the center of the screen. The camera can be placed anywhere in 3D space either inside or outside the pyramid. To find the current target position select the **POints** option.

CAmera/TArget/Distance/POints/PAn/Zoom/TWist/CLip/
   Hide/Off/Undo/<eXit>: **PO**
Enter target point < >: **1300,1300,0**
Enter camera point < >: **1300,900,237**

This puts the target at the center of the pyramid base and the camera in front of and looking down on the eye (Figure 8.23a). That was a static type of operation. To use the dynamic view select the "CAmera" option. This allows you to specify the angle from the XY plane (the base ABCD of the pyramid). A positive angle puts the camera above the target, a negative angle below. A plan view can be generated by using an angle of 90 degrees. You are then asked to put in a camera direction angle relative to the X axis (line AB). This angle rotates the camera in a horizontal plane while keeping it focussed on the target point.

When you select the **CA** option you are prompted for new angles with the current values as defaults. You can enter the angle by typing a value or by moving the cross hairs on the screen. Move the cursor up and down to give the angle from the XY plane and right to left for the angle in the XY plane. The pyramid will be rotated and shown ghosted, in preview mode. The preview image is updated continuously. The speed of the update will depend on the number of objects in the dview selection set. Make sure that SNAP is OFF or the action may appear jumpy (Use ^**B**).

CAmera/.../PAn/Zoom/TWist/CLip/Hide/Off/Undo/<eXit>: **CA**
Toggle angle in/Enter angle from XY plane <30.65>: **T**

Pressing T toggles the angle to "from X axis". The angle in the toolbar will now respond to left–right movements of the mouse.

Toggle angle from/Enter angle from XY plane from X axis < −90>: **-35**
Toggle angle in/Enter angle from XY plane <30.65>: **5**

If you use the mouse to pick values both values will be input simultaneously based on the cross hairs' position. Using the toggle and the mouse with a steady hand is advised. Try to restrict the mouse movements firstly to left–right and then to up–down. The screen acts like a scale for the angles e.g. the top is $+90°$ and bottom of the screen in $−90°$ with zero being at the center. When a suitable view is found, then press the mouse button. If one value has been keyed in then the mouse position will only be used for the remaining angle.

The TArget option works very like CAmera except that it is the target point that moves relative to the camera position. It is not very useful. Dview's PAn and Zoom options work similarly to the normal commands. The zoom is, however, a restricted version of the normal command. It only allows you to change the magnification, similar to the ZOOM with Scale(X). You get a slider bar so you can see the effect before picking a scale factor. A scale factor less than 1 reduces the size and greater than 1 increases it. If you increase the scale too much the object might disappear. It hasn't gone anywhere, it's just that you are zoomed in on a single brick. Zoom back out to see the whole thing.

CAmera/TArget/.../Pan/Zoom/.../Off/Undo/<eXit>: **Z**
Adjust zoom scale factor <1>: Use slider bar and press <**ENTER**>

TWist lets you rotate the view in the plane of the screen about the target point. This has the effect of rotating the camera on the line of sight. A rubber band appears from the target to the cursor cross-hairs and shows the current angle of twist. The angle is zero when the rubber band is horizontal and to the right. The camera is upright when the twist angle is zero, upside-down when the angle is 180 degrees and on its side for 90 degrees. The twist angle is an additional setting and does not affect the camera or target positions.

CAmera/.../Zoom/TWist/CLip/Hide/Off/Undo/<eXit>: **TW**
New view twist <0.00>: Move cursor around and press <**ENTER**>

One of the more useful features of the DVIEW command is the ability to generate cut-away images with the CLip option. This allows you to specify planes in front of and behind the target to cut through the object. Nothing between the camera and the front plane will be displayed. Similarly nothing behind the back plane is shown. This can be used to eliminate unecessary foreground and background detail or to generate a cut-away view.

CAmera/.../Zoom/TWist/CLip/Hide/Off/Undo/<eXit>: **CL**
Back/Front/<Off>: **F**
Eye/ON/OFF/<Distance from target> <464.94>: **215**

A front distance of 464.94 places the plane at the camera in this instance while 0 would put it at the target point. Thus, 215 is between the camera and target, within the limits of the pyramid. The "Eye" option places the front plane at the camera point. This is useful for perspective views when clipping cannot be turned off. In normal dynamic viewing you can turn the front clip ON and OFF.

To get a better view of the front clip use the Hide option (Figure 8.23d). This does a hidden line removal just like the HIDE command. Then clip a piece off the back and remove the hidden lines.

CAmera/.../Zoom/TWist/CLip/Hide/Off/Undo/<eXit>: **H**
CAmera/.../Zoom/TWist/CLip/Hide/Off/Undo/<eXit>: **CL**
Back/Front/<Off>: **B**
ON/OFF/<Distance from target> <−149.93>: **−300**
CAmera/.../Zoom/TWist/CLip/Hide/Off/Undo/<eXit>: **H**

The minus indicates that the plane is to be behind the target. These clipping planes will remain in effect until CLip is turned off.

Getting things in perspective

To get a realistic view of objects you can generate a perspective view. When objects are in perspective the ones nearer the camera appear bigger than those further away. You can control the perspective view by choosing the "Distance" option from DVIEW. You specify the distance from the camera to the target point and AutoCAD calculates the appropriate sizes of the objects. Again, a

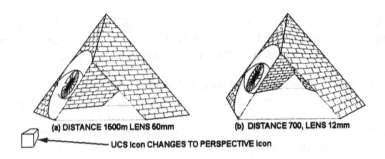

Figure 8.24 Perspective views

slider bar is available to input the distance via the mouse, and the status line gives a read-out of the current slider bar position.

> CAmera/TArget/Distance/.../CLip/Hide/Off/Undo/<eXit>: **D**
> New camera/target distance <464.94>: **2500**

WARNING! The UCS icon should change to a perspective view of rectangular block (Figure 8.24). Most of AutoCAD's commands do not work on perspective views, so keep your eye out for this icon.

The dview zoom operates slightly differently when a perspective view is displayed. Instead of a scale factor AutoCAD asks for a camera lens length. The default is 50mm which is the standard lens focal length for most cameras. Making the lens length longer is like using a telephoto lens and magnifies the image. A shorter lens length simulates a wide angle camera lens which accentuates the perspective (Figure 8.24). The slider bar gives the zooms in multiples of the current lens length.

> CAmera/TArget/Distance/.../Zoom/.../Hide/Off/Undo/<eXit>: **Z**
> Adjust lenslength <50.000mm>: **12**

To get an interesting fish-eye lens effect, shorten the perspective distance and do another hide.

> CAmera/TArget/Distance/.../Zoom/.../Hide/Off/Undo/<eXit>: **D**
> New camera/target distance <1000>: **700**
> CAmera/TArget/Distance/.../Zoom/.../Hide/Off/Undo/<eXit>: **H**

To round off the DVIEW command, the "Off" option turns the perspective viewing off. "Undo" goes back to the previous view and "eXit" leaves the DVIEW command. The display retains the DVIEW settings on exit. It is dangerous to remain in a perspective view so switch it off. If there are any dviews that you wish to retain then exit the command and save them as a named view. The named view will retain all the information including any perspective setting and clipping planes.

CAmera/TArget/Distance/.../Zoom/.../Hide/Off/Undo/<eXit>: **Off**
CAmera/TArget/Distance/.../Zoom/.../Hide/Off/Undo/<eXit>: **X**
Command: **QSAVE**

Two more pyramids

Of course, the complex at Giza consists of 3 pyramids with Cheops' being the largest. No Express visit would be complete without including the two lesser, but still magnificent, structures.

Switch to the plan view in the World coordinate system by picking **View/3D Viewpoint** followed by **Plan View/World UCS**. Then do a Zoom- all. Bring up the Surfaces toolbar using **View/Toolbars** and pick the **Pyramid** button and draw the two structures shown in Figure 8.25.

Command: ai_pyramid	
First base point: **400,650,0**	(A)
Second base point: **@400,0,0**	(B)
Third base point: **@0,400,0**	(C)
Tetrahedron/<Fourth base point>: **@-400,0,0**	(D)
Ridge/Top/<Apex point>: **600,700,250**	(E)
Command: **Copy**	
Select objects: **last** 1 found	
Select object: **<ENTER>**	
<Base point or displacement>/Multiple: **400,650,0**	(A)
Second point of displacement: **1900,1550,0**	(F)

The pyramids in Figure 8.25 should now be in plan view. The view in Figure 8.25(b) was obtained by trial and error with the VPOINT command's "Rotate" option . To get the impression of a human viewpoint, the rotation from the XY plane is very small.

Command: **VPOINT**
Rotate/<Viewpoint>: **R**
Enter angle in XY plane from X axis <270>: **230**
Enter angle from XY plane <90>: **3**
Command: **HIDE**
Command: **QSAVE**

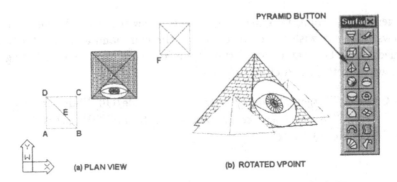

Figure 8.25 3 Pyramids

That concludes the AutoCAD Express stop in Egypt. It doesn't conclude the exploration of AutoCAD's third dimension. The next section covers more exciting features.

AutoCAD's 3D box of tricks

The Surfaces toolbar has a load of special commands for generating 3D objects. You have already used the 3DFACE command and a Pyramid objects. You can also generate smooth surfaces and 3D polylines. The toolbar contains all the commands except for 3D polyline which is on the Draw menu.

The 3DPOLY command allows you to create a polyline in 3D space. The PLINE command is restricted to 2D. Points can be specified in the same way as for drawing a 3D line but the use of 3DPOLY is restricted to straight line segments. The line width is zero and cannot be changed. Neither can you draw a 3D polyarc. You can use PEDIT on a 3D polyline to change any of the vertices or to fit a spline curve to the points.

Start up a new drawing with limits (0,0) to (1200,850). You will use this to draw examples of the 3D constructions. Pick **File/New** and, **Start from scratch** with metric units. It will be called **EXP-3D**. Then draw a spiral.

Command: **LIMITS**
Reset Model space limits
ON/OFF/<Lower left corner> <0.00,0.00>: <**ENTER**>
Upper right corner <420.00,297.00>: **1200,850**
Command: **ZOOM**
All/.../Window/<Realtime>: **A**
Command: **3DPOLY** or Pick **Draw/3D Polyline**
From point: **100,100,0**   (A)

Close/Undo/<Endpoint of line>: **@125,0,25**   (B)
Close/Undo/<Endpoint of line>: **@0,125,25**   (C)
Close/Undo/<Endpoint of line>: **@−130,0,25**   (D)
Close/Undo/<Endpoint of line>: **@0,−130,25**   (E)
Close/Undo/<Endpoint of line>: **@135,0,25**   (F)
Close/Undo/<Endpoint of line>: **@0,135,25**   (G)
Close/Undo/<Endpoint of line>: **<ENTER>**

Note that the 3D polyline cannot be assigned a width. It can be edited in other ways using PEDIT. For example make this polyline into a spiral (Figure 8.26).

Command: **PEDIT**
Select polyline: **LAST**
Close/Edit vertex/Spline curve/Decurve/Undo/eXit <X>: **S**
Close/Edit vertex/Spline curve/Decurve/Undo/eXit <X>: **<ENTER>**

PEDIT recognized that the object is a 3D polyline and offers only the relevant editing options. The accuracy of the fit is controlled by a system variable "SPLINESEGS". If the system variable "SPLFRAME" is non-zero then the original polyline is also shown. This gives an idea of how accurately the curve fits the points. Do this and change the view point to see the graceful spiral. Non-zero SPLFRAME also shows up any invisible 3DFACE edges.

Command: **SPLFRAME**
New value for SPLFRAME <0>: **1**
Command: **VPOINT**
Rotate/<View point>: <0.00,0.00,1.00>: **1,−1,1**

Meshes

You can build up the surface of an object by using lots of 3DFACEs but it could take a long time. A slight improvement is to use the 3DMESH command but this still requires you to input the coordinates of each vertex. If the object has a regular shape then the mesh can be generated from various control lines using either EDGESURF, REVSURF, RULESURF or TABSURF.

If you do not want to input all the vertices of a mesh you can use a little routine that is hidden in an old dialog box. Pick **Draw/Surfaces** followed by **3D Surfaces....** This brings up the 3D Objects dialog from which you can pick **Mesh** and pick **OK**. You then specify the number of vertices. These are given in the length and breadth directions (M and N, respectively). First, set SPLFRAME to zero again.

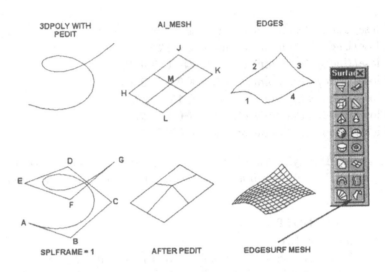

Figure 8.26 3D spiral and meshes

Command: **SPLFRAME**
New value for SPLFRAME <1>: **0**
Pick **Mesh** from the 3D Objects dialog box
Command: ai_3dmesh
First corner: **300,100,50**   (H)
Second corner: **300,250,70**   (J)
Third corner: **400,250,70**   (K)
Fourth corner: **400,100,50**   (L)
Mesh M size: **3**
Mesh N size: **3**   (This will make four faces in the mesh.)

You can use PEDIT on 3DMESHes as well (Figure 8.26). The edit options
are different when a mesh is selected. You can smooth the mesh, close it in
either M or N directions or change individual vertices. Use PEDIT to raise the
middle vertex (1,1), M, by 20 units. Do a **Zoom All** if you can't see the whole
mesh.

Command: **PEDIT**
Select polyline: **LAST**
Edit vertex/Smooth surface/Desmooth/Mclose/Nclose/Undo/
  eXit <X>: **Edit**
Vertex (0,0). Next/Previous/Left/Right/Up/Down/Move/REgen/
  eXit <N>: **U**

The Up/Down refers to movements in the M direction and Left/Right the N direction. When you get to the desired vertex you can "Move" it.

Vertex (1,0). Next/Previous/Left/Right/Up/Down/Move/REgen/ eXit <U>: **R**

Vertex (1,1). Next/Previous/Left/Right/Up/Down/Move/REgen/ eXit <R>: **M**

Enter new location: **@0,0,20**

Vertex (1,1). Next/Previous/Left/Right/Up/Down/Move/REgen/ eXit <R>: **X**

Edit vertex/Smooth surface/Desmooth/Mclose/Nclose/Undo/ eXit <X>: **<ENTER>**

Note that the numbering of the vertices starts at (0,0) and so (1,1) is the second across and second up. Smoothing is only relevant when there are more than two faces in one of the directions. The command AI_MESH is a big improvement on the old 3DMESH command. It does have a silly name though. With a slightly different approach, the same effect as ai_mesh can be achieved using EDGESURF as demonstrated in the next section.

Generated surfaces

The 3DPOLY is quite good for defining the edges of a surface. Once the edges are known, EDGESURF, RULESURF, TABSURF and REVSURF can be used to fill in the surface. TABSURF requires one edge and an extrusion direction, REVSURF needs a profile edge and an axis of revolution. RULESURF is defined by two edges, while EDGESURF is the most complicated, requiring four edge curves.

EDGESURF works by interpolating a Coons surface patch between four curves. The Coons patch is a mathematical technique using two cubic equations. The edges can be made up of lines, arcs, or open polylines and must touch at their end points. Use 3DPOLY to create four connected curves and then pick the Edge Surface button or **Draw/Surfaces** and **Edge Surface** from the menu bar. This executes the command EDGESURF to make a mesh like the one in Figure 8.26.

Command: **EDGESURF**
Select edge 1: Pick the first edge curve.
Select edge 2: Pick the second edge curve.
Select edge 3: Pick the third.
Select edge 4: Pick the fourth.

This generates a polygon mesh which can be edited in the same way as the previous mesh. The vertices are numbered with the M direction along the first

edge curve. The (0,0) vertex will be at the end point of the first edge nearest to
the pick point used to select it. The number of faces that are created depends
on the values of the two system variables, SURFTAB1 for the M direction and
SURFTAB2 for N. The defaults for these are 6 each and they can be changed
by typing Surftb1 or Surftb2 at the command prompt. Values of 12 each were
used for Figure 8.26.

You must have four edges to define an EDGESURF mesh. If the shape
requires only three curves then use BREAK to split one of the sides in two.

If some of the sides can be defined by straight lines or regular shapes such
as arcs and circles then the commands TABSURF, RULESURF or REVSURF
may be more appropriate (see Figure 8.27).

The Tabulated Surface or TABSURF is good for extruding objects in 3D
space. It gives an effect similar to setting an object THICKNESS but is more
general as the extrusion direction is controlable. To generate a TABulated
SURFace you require some object, called path curve in AutoCAD, to extrude
and a line defining the direction of extrusion. In descriptive geometry jargon
you need a *directrix* object and a *generatrix* vector. To make a leaning tower
draw a circle in the WCS plan and a line.

> Command: **CIRCLE**
> 3P/2P/TTR/<Center point>: **500,300,0**
> Diameter/<Radius>: **50**
> Command: **LINE**
> From point: **570,300,0**
> To point: **@0,40,100**
> To point: <**ENTER**>
> Command: **TABSURF** or Pick **Tabulated Surface** from the menu.
> Select path curve: Pick the circle.
> Select direction vector: Pick the line near the bottom.

This generates an open ended leaning tower. It is not cylindrical though, since
the direction vector is not perpendicular to the plane of the circle. The extrusion direction depends on the location of the point picked on the direction
vector. Pick near the bottom and the extrusion is upwards towards the other
end of the line.

A Ruled Surface is more general than the tabulated surface. You specify
two boundary edges and RULESURF joins them together with straight lines
to form a polygon mesh. You can use open or closed 2D and 3D polylines,
circles, arcs, lines and points. However you cannot mix a closed object such
as a circle with an open object such as a line. Points can be used with either
open or closed paths. Make a surface from an arc to a line. Remember, ARCS,
like circles are drawn in the plan of the current UCS.

> Command: **ARC**

Center/<Start point>: **800,300,0**
Center/End/<Second point>: **C**
Center: **750,300**   (You must give a 2D point.)
Angle/Length of chord/<End point>: **A**
Included angle: **270**
Command: **LINE**
From point: **800,500,0**
To point: **@−220,0,30**
To point: **<ENTER>**
Command: **RULESURF** or pick **Ruled Surface** from the menu.
Select first defining curve: Pick the arc near the start point (A)
Select second defining curve: Pick the line near B.

The two end points nearest the places where the curves are picked define the starting vertices of the mesh. If you pick one of the curves near the wrong end the surface will be twisted. If this happens use the UNDO command and try again, picking the points at the correct ends of the curves. The number of divisions for both TABSURF and RULESURF is determined by the SURFTAB1 system variable.

The final surface generator in the 3D box of tricks is the Revolved Surface or REVSURF. This produces a surface of revolution from a definition path and an axis to rotate it around. Common surfaces of revolution include wine goblets, spheres, torus shapes and hyperboloids. REVSURF allows you to make either closed surfaces or open ones by controlling the angle of rotation. To make a part sphere draw a circle with an axis along one diameter.

Command: **CIRCLE**
3P/2P/TTR/<Center point>: **1000,300,0**
Diameter/<Radius>: **100**
Command: **LINE**
From point: **860,300,0**
To point: **@280,0**
To point: **<ENTER>**
Command: **REVSURF** or pick **Surface of Revolution** from the menu.
Select path curve: Pick the circle.
Select axis of revolution: Pick the line near the left hand end.
Start angle <0>: **<ENTER>**
Included angle (+ccw −cw) <Full circle>: **90**

The positive direction of the axis depends on where the line defining the axis is picked. Angles are positive in the anti-clockwise direction as you look from the picked point to the furthest end point.

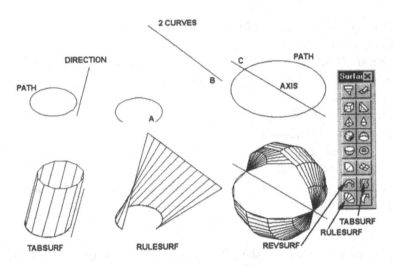

Figure 8.27 3D generated surfaces

3D objects

More routines can be obtained by picking the other buttons on the Surfaces toolbar shown in Figure 8.27. These allow you to easily produce a Box, Wedge, Pyramid, Cone, Sphere, Dome, Dish and Torus. The actual commands that are executed when these buttons are selected are "ai_box", "ai_wedge" etc. They must not be confused with the 3D solid commands of similar name. BOX and Wedge produce solids while ai_box and ai_wedge produce surfaces. The number of divisions in the sphere, hemi-sphere, torus and cone is controlled by SURFTAB1 and SURFTAB2. The larger these numbers, the smoother the surfaces will appear. All of these commands produce 3d meshes which can be edited usin PEDIT.

3D Solids

The difference between and AutoCAD solid and a 3D surface is the same as the difference between a solid wooden block and a hollow wooden box. From the outside, they look the same but inside they are different. AutoCAD solids are special objects that have mass properties, centers of gravity etc. The basic solid shapes, box, sphere, cylinder, etc. are known as "primitives". More complicated solids can be formed by adding (union) or subtracting solids from one another. You can also obtain the intersection volumes of solids. These are called "Boolean" operations after the 19th century mathematician.

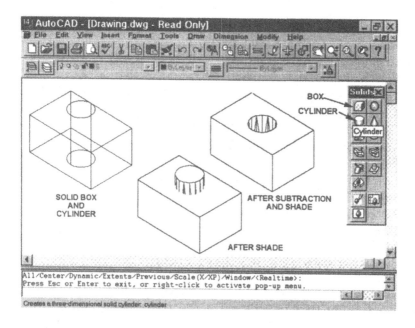

Figure 8.28 3D Solid

Solid modelling is a special form of CAD which itself merits a whole book on the various techniques needed. As such it is beyond the scope of this short book. However, to give you a flavor for the topic and to expose some of the new facilities Figure 8.28 shows a short example.

Pick **View/Toolbars** and select the **Solids** toolbar. Using the same drawing file as above create a box and then subtract a cylinder from it to make a hole.

Command: **Box**
Center/<Corner of box> <0,0,0>: **100,500,0**
Cube/Length/<other corner> **@300,200**
Height: **150**
Command: **Cylinder**
Elliptical/<center point> <0,0,0>: **250,600,0**
Diameter/<Radius>: **50**
Center of other end/<Height>: **200**

If you select an isometric view, the cylinder and box don't look that great. Even the hide command doesn't seem to help. You get a better picture if you SHADE the objects. Then, do the Boolean subtract to create the hole in the box.

```
Command: VPOINT
Rotate/<View point> <0,0,1>:1,-1,1
Command: SHADE
Shading complete.
Command: REGEN                    (Needed before the subtraction)
Command: SUBTRACT
Select solids and region to subtract from . . .
Select object: pick the rectangular box
Select objects: <ENTER>
Select solids and regions to subtract. . .
Select objects: pick the cylinder
Select objects: <ENTER>
Command: SHADE
Shading complete.
Command: SAVE give the name EXP-3D
```

Figure 8.28(c) shows the resultant solid after the SHADE command. Your object might appear in reverse colors. Shading depends on the setting of a variable called SHADEDGE. The image here was produced with SHADEDGE equal to 2. More elaborate shading can be done using AutoCAD's Render facility described in the next chapter. To really appreciate that the object is solid you can obtain its mass properties with the MASSPROP command. This can be found on the **Tools/Inquiry** menu.

Use of 3D solids should be reserved for situations where the solidity is essential. There is a heavy overhead in both drawing technique and processor power when using solid modelling. Very often a 3D surface will suffice.

Summary

This chapter has covered the use of AutoCAD in isometric projection, 2.5D and full 3D. You have also encountered the display features of user coordinate systems and view points. The UCS and MVIEW commands can be used in 2D drafting as well as 3D. Dynamic viewing is a powerful aid to visualizing 3D spatial relationships.

The UCS facility is the most important tool in 3D computer aided drafting. Many objects are 2D (eg arcs, circles) and can only be drawn in plan. To draw a sloping circle you have to create a coordinate system so that the plane of the circle is the same as the plane of the UCS.

Using 3D CAD involves an extra level of difficulty above 2D drafting and requires much more discipline. You must keep track of where objects are and also what coordinate system and view point is being used. Vigilance helps to prevent troublesome errors caused by optical illusions. 3D surfaces can create the same shapes as 3D solids but do not have the mass properties of the solid objects.

You should now be able to:

Draw lines and circles in isometric projection.
Create objects with different elevations and thicknesses.
Use hidden line removal.
Set up multiple viewports.
View 2.5D and 3D objects from different view points.
Define new coordinate systems in 3D space.
Draw objects in 3D space.
Run dynamic visualisations and create perspective views.
Generate 3D surfaces.
Create compound 3D solids.

Chapter 9 THE HARDCOPY – PRINTING AND PLOTTING

General

The main purpose of producing drawings with AutoCAD is to communicate graphical information. Even as the twentieth century makes way for the twenty first, the primary medium for such communication is with pictures on paper. Paper drawings are very user friendly, they are easy to read and transport and also provide a useful framework for rough work and checking. With this in mind the current chapter is devoted to methods of translating the digital information in the AutoCAD drawing into marks on paper.

A new order of activity is involved in producing plots and prints. That is you have to control another piece of equipment, be it printer, plotter or both. According to Murphy's Law this extra complexity inevitably leads to more things that will go wrong. To avoid the heartache and frustration associated with peripheral blues, stay calm and follow the guidelines laid out below.

Unless you have access to very expensive printing equipment the generation of a hardcopy of a drawing takes time. The cheaper your plotter the longer it will take. In general you will not want to have to reproduce prints of large drawing files too often and so the aim is to get it right first time, if at all possible.

You will produce copies of some drawings from earlier chapters. Make sure that the drawing files BALLOON.DWG, from Chapter 4, GLAND.DWG, from Chapter 7, and EXP-GIZA from Chapter 8 are handy. You will also create a standard title block in AutoCAD's paper space and generate a multiple view plot of the all seeing pyramid. If you don't have these files you can improvise with some other simple drawings. Avoid printing bigger drawings until you have more experience. It is assumed that AutoCAD has been configured correctly for your printer and plotter. If you are not sure if this has been done refer to Appendix A or the *AutoCAD Installation Guide*.

Printing and plotting

Printers and plotters

Printing technology has developed rapidly with laser and bubble jet devices now the most common printers. The quality of the printed output depends on the resolution capability of the device. The printer produces the picture by converting all the graphics into a series of dots which are then inscribed on the paper. The crucial statistic for assessing resolution is the number of *dots per inch* (dpi) capability of the printer. The higher the dpi, the better the picture quality and usually the slower the print speed. The quality is most noticeable when plotting arcs and other curved entities and shallow sloping lines.

Printer memory is another key factor in improving productivity. The old adage, *the more the merrier* holds true here. Extra printer memory will speed up the transfer of the drawing from computer to printer and will allow higher resolution output.

While traditional engineering drawings were done on large sheets (typically A0 size) , the quality of CAD output means that most working drawings are now on smaller sheets (A3 or A4). A3 and A4 printers are much cheaper than the big boys.

If you are a seasoned CAD professional you will remember the old fashioned pen plotter. Such devices used mechanical arms and rails to move real pens across the page producing very high quality (and very slow) output. Since AutoCAD has come out of the plotter era many of its old commands (e.g. the PLOT command) have been updated to behave just like PRINT in any other Windows application. Thus, the two commands "PLOT" and "PRINT" are equivalent in AutoCAD.

HAZARD WARNING! As a general precaution, before issuing any print or plot instructions from within the AutoCAD editor you should save the drawing. This is good CAD practice as a malfunction of the peripheral can cause the computer to hang up. If you have to reboot the system you will lose everything drawn since the last SAVE command.

Printing the GLAND drawing

For the first printout, you need to start up AutoCAD and OPEN the Gland drawing from Chapter 7. To do this select **Open a Drawing** button from the Start Up dialog. If AutoCAD is already running then pick **File/Open** from the menu bar. Then click on **gland.dwg** from the list of file names. When the file has been loaded thaw the frozen layers, switch to the GLAND layer

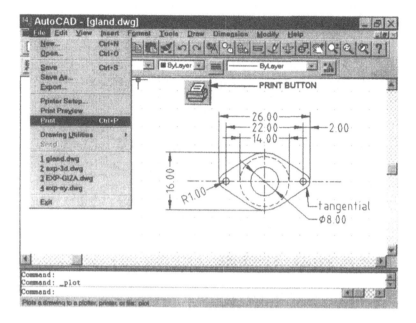

Figure 9.1 Plotting the gland

and freeze the POLYGON layer. Then pick **Print...** from the **File** menu (Figure 9.1).

The first thing to do when the Print/Plot Configuration dialog box appears is to check which output device you are going to use. Pick the button marked **Device and Default Selection** in the top left part of the dialog screen (Figure 9.2). This will give a list of the available printers and plotters. Pick the device of your choice. In this example, I have chosen the Windows System Printer. Using the Default System Printer means that you can use the Standard Windows methods for selecting and configuring the printer.

No dialog box? If the dialog box in Figure 9.2 did not appear but instead you got some text prompts in the Command area of the screen, here's what to do. The dialog box is controlled by the AutoCAD system variable called **CMDDIA**. If this is zero the dialog will not appear. Use **Esc** to cancel the PLOT command and type CMDDIA at the command prompt and give it the value **1**. Then pick **File/Print...** once more.

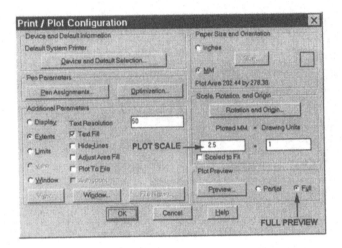

Figure 9.2 Print/Plot configuration dialog box

Altering the plot parameters

Remember, the limits were set to (0,0) and (65,45) in Chapter 7. However, as the actual drawing doesn't quite fill the limits we will plot the "Extents" of the drawing. Specifying **Extents** in the Additional Parameters section of the plot dialog box shown in Figure 9.2 causes AutoCAD to calculate a window just big enough to fit in all the drawings entities. The main alternatives to choosing extents are using the drawing limits or the current display. The Window button behaves as with the ZOOM command and allows you to plot part of the drawing.

Text fill will ensure that the True type fonts will come out solidly. The text resolution to be used for True type fonts. The "Hide Lines" is only applicable for 3D drawings and is dealt with later. The "Adjust Area Fill" is used when the edges of the solids and polylines must be positioned very accurately (e.g for printed circuit boards). It saves AutoCAD some calculation time if the adjustment is not requested.

In some cases, it is desirable to generate plot files which can be sent to the printer later. For example, if you had a large number of drawings to print or if the printer was not available, you could still generate the plot files, store them on disk and print them later. Here, we do not want to generate a plot file.

When using the Windows System Printer, the page size is defined in Windows. As such AutoCAD does not allow you to change it and offers the paper size reported to it by the Windows printer properties set up. If you want to

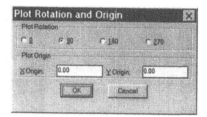

Figure 9.3 Rotating the plot

select a different page size then use your Printer properties dialog. This can be found by picking the Device & Default Selection button followed by the Change button towards the bottom of the subsequent dialog box. Alternatively, you can pick **Start** followed by **Settings** and **Printers** from the Windows Start menu.

Here, AutoCAD reports that the plot area is 202mm by 278mm. This is despite the fact that the printer is set up for A4 paper (210mm by 297mm). AutoCAD deducts a small margin from the size since the printer cannot print right up to the edge. In general, the actual area available for printing is device dependent. You will need to get to know your printers to discover their capabilities.

Scales and Preview

The observant reader will have noticed that the printer paper size is given in Portrait orientation. However, the AutoCAD screenand the gland drawing are in Landscape. Pick the **Rotation and Origin** button and specify a rotation of **90**.

If the Scaled to Fit box is checked, AutoCAD calculates a scale factor to fit the plot into the maximum area. This usually results in silly scales such as 278 plotted mm = 58 drawing units. Engineers and designers are used to reading drawings at specific scales (2:1, 1:1, 1:5 etc). Faced with a scale of 5.129:1, difficulties arise. Thus, here scale will be prescribed by the user. Make sure that the **Scaled to Fit** option is not checked. Then input **2.5** for the plotted mm and 1 drawing unit as shown in Figure 9.2. Thus if each drawing unit represented 1mm in real life the plot will be 2.5:1 ie two and a half times bigger than reality.

Before executing the plot it's best to get a preview of what the output will look like. The Plot Preview button is towards the lower right of the dialog box (Figure 9.2). There are two options, Partial or Full. A partial preview will show only the limits of the plot in relation to the paper. This is useful for

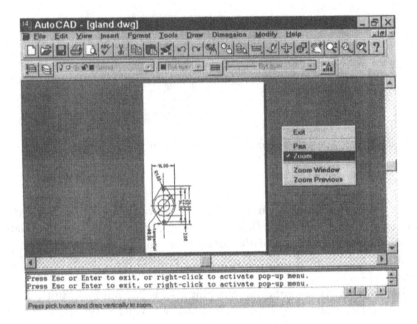

Figure 9.4 Previewing the plot

checking that the orientation and origin are sensible. However, if parts of your drawing exceed the paper size you will not know from this. A full preview will show exactly what the plot will look like. Pick the **Full** button and then pick **Preview**. Your display should then resemble Figure 9.4. The Pan and Zoom work like Zoom/Realtime and the cursor becomes the familiar magnifying glass or hand. Right click the mouse to get the **Pan and Zoom** menu.

On the preview shown in Figure 9.4, the gland is too near the bottom left corner of the page as it appears on the screen. To center it on the page press **Esc** to exit the prieview and pick the **Plot and Origin** button again. Give a value of **50** for the X origin and **60** for the Y origin and pick **OK**. Note that the Y origin shift moves the gland up the *screen* in the prieview. Check the preview once more and alter the origin if needed. If all is well, then press **Esc** to exit preview and pick **OK** in the Print/Plot Configuration dialog box. AutoCAD process the file and send it to the printer. When this is done you will get the "Plot complete" message. Note that the UCS icon and GRID dots are not included in the print out.

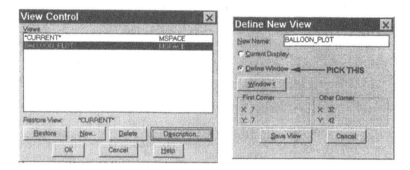

Figure 9.5 View creation

Saving a VIEW

In this part of the exercise, you will generate a number of multicolored plots of the Balloon drawing from Chapter 4. It is often necessary to produce a number of different plots of various parts of a drawing. You might want a large scale view of some detail as well as a general layout. You can explore the drawing using ZOOM, PAN and VPOINT or DVIEW. Once you are happy with the display it can be stored as an AutoCAD VIEW. Such views can be quickly retrieved for redisplay or plotting. This saves you from having to remember what ZOOMs, clipping planes and other operations were used to get the desired effect. In this section you will create two VIEWs of the BALLOON drawing.

To open the drawing file pick **File/Open** followed by **balloon.dwg** from the file list. Named views are created by picking **View/Named View** or the icon from the Viewpoint toolbar. In the View Control screen (Figure 9.5) pick **New**. This allows you to define a new view. Give the name as **BAL-LOON_PLOT** and pick the circle beside "Define window". Then pick the **Window** button. This returns you to the graphics screen to pick a window or type the coordinates.

Command: _ddview
First corner: **7,7**
Other corner: **32,42**

Finally, pick **Save View** to add it to the list shown in the view control box.

This allows you to save a rectangular section of the display. Of the other options, "Description" gives details of the settings associated with the view; "Delete" allows you to delete a stored view (but not the one named "*CUR-RENT*"; "Restore" causes a named view to be displayed on the graphics screen. In model space, if you have more than one viewport on the screen only

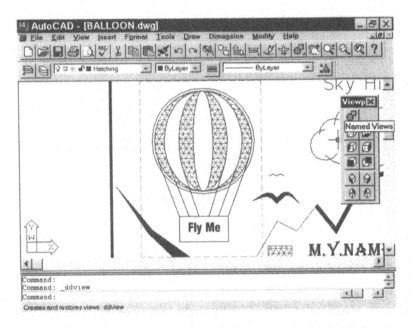

Figure 9.6 BALLOON_PLOT view

the active one can be stored as a VIEW. The rules for view names are the same as for layers, up to 31 numbers or letters but no spaces or full stops. The "MSPACE" after the view name indicates that is was created in model space.

Note that this is a tall, thin VIEW. When it is displayed on the screen, parts of the drawing to the left and right of the view's window may be displayed. However, when it is plotted only that portion within the window defined by (7,7) to (32,42) will be drawn. To see the view (Figure 9.6), pick the line:

BALLOON_PLOT   MSPACE

followed by the **Restore** button and the **OK**. The dashed lines indicate the limits of the named view. They will not appear on your screen. When this view is eventually printed, only the parts within the dashed lines will be output.

To generate the second view that is plotted in Figure 9.9, ZOOM in to the lower right corner and use the "Current Display" option in the Define New View dialog box.

Command: **ZOOM**
All/.../Window/<Realtime>: **W**
First corner: **30,0**

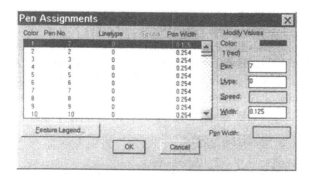

Figure 9.7 Setting print colors

Other corner: **65,25**

Command: **DDVIEW** or pick View/Named View...

When the View Control dialog box (Figure 9.5) appears pick **New...**, give the new name as **MOUNTAIN_PLOT**, pick the **Current Display** radio button followed by the **Save View** button. To exit the View Control pick **OK**. You now have two suitable views for plotting.

Multi-colored plotting

The ability to do multi-colored prints obviously depends on you printer. Most laser printers produce either black and white output or greyscale. Many colour devices are also available. With most pen plotters devices you can have more than one pen and so vary the colors or line thicknesses. You can also vary the pen speed on some plotters to get better quality lines.

Remember to save the drawing before plotting. Use the quick save command, picking **File/Save**. To generate a plot of the mountain view pick **File/Print...** from the menu bar. Check the pen settings by picking the **Pen Assignments** button from the Plot Configuration screen. The dialog box shown in Figure 9.7 should then appear.

The first column indicates the AutoCAD color number, 1=red, 7=black etc. Next comes the pen numbers. This is a throw back to the days when all color output was done on pen plotters. The color mapping depended on how you loaded you plotter and how many pens your had. Generally, for color printers there is a one to one mapping between the AutoCAD color and the printer color and you do not have to modify the pen numbers. If you are using a printer with only greyscale capability then being able to alter the "pen" number becomes important.

With some laser printers, yellow lines and other colors will come out very faint if not almost invisible. First you select the line or lines you wish to modify from the list and then use the options on the right. Note that AutoCAD has as many colors as your display allows. In Figure 9.7 color number 1, red, has been reset to use pen number 7. Pen number 7 is the the same as AutoCAD color 7 i.e. black. This is done by picking the line:

1   1   0   0.254

Then using the "Modify Values" area on the right, move to the **Pen:** field and key in **7**. You can also change the plotter linetype, pen speed and width setting. The linetypes here are not AutoCAD linetypes but plotter ones. It is not recommended to mix AutoCAD and plotter linetypes so it is best to leave Ltype at 0. The pen speed option would be useful if it worked. With my old pen plotter device it doesn't. Thus it is safest to set the pen speeds using the plotter's own console. Some pens operate better at slower speeds eg drafting pens. The speed is irrelevant for laser printers.

Finally the line width can be important for both printers and plotters. The thicknesses in Figure 9.7 are in mm. The value shown is 0.254mm which happens to be 100th of an inch which is actually quite thick so it has been changed to **0.125** for color 1. The other colors and widths are left unchanged. For pen plotters, AutoCAD calculates the number of passes of the pen that are required to fill in a thick polyline. If the specified pen width is larger than the actual pen width solid objects will appear striped rather than filled.

Productivity tip. As mentioned above, it can be very useful to manipulate color mappings for greyscale laser printers. You can use colors in AutoCAD and then map all these to black with different line widths. It is possible to select all the colors by dragging the mouse down the list with the left button depressed. You can then apply pen number 7 to them all if you wish.

Plotting a view

Having made all the modifications you require to the pen assignments, pick **OK** to get back to the Print/Plot Configuration screen. Now pick **Device and Default Selection** and in the new dialog box pick **Save** under "Complete (PC2)" shown in Figure 9.8. Accept the default file name, "BALLOON.PC2" and **Save**. This stores all the pen assignments and other device independent print settings. The Replace button alows you to load the saved pen assignments in other drawings. Pick **OK** to get back once more to the Print/Plot dialog.

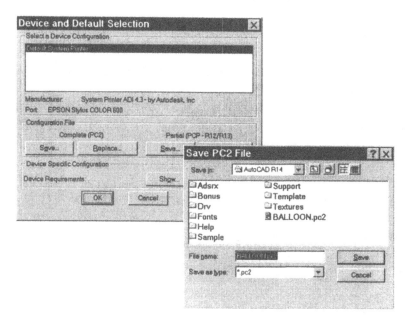

Figure 9.8 Saving pen assignments

Now pick the **View...** button in the lower left corner. This gives a window with all the views defined in the drawing. Pick **MOUNTAIN_PLOT** followed by **OK**.

Postscript output

An increasing number of printers and electrostatic plotters accept the Postscript command language by Adobe. Postscript is a powerful plotting language. It is hardware-independent and the industry standard for plotting graphics and desk-top publishing.

For report generation, it is often necessary to combine AutoCAD drawings with wordprocessed documents. One powerful way to do this is to export the information to a postscript file using the **PSOUT** command. This is found by picking **File/Export....** This gives the Export Data dialog shown in Figure 9.10. Select **Encapsulated PS (*.eps)** from the "Save as type" list and then pick **Options**. Choose **A4** paper size and pick **View**. From the View Name list pick **BALLOON_PLOT** followed by **OK**. Back at the Export Data dialog pick **Save** to create BALLOON.EPS.

Figure 9.9 Mountain plot output

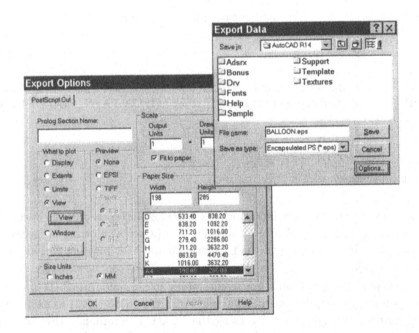

Figure 9.10 Export data

Figure 9.11 Balloon Postscript plot

This file can be imported to many other packages or sent directly to a postscript printer. Note that the image plotted in Figure 9.11 is restricted to the limits of the view shown dashed in Figure 9.6.

The Export Data dialog allows you to save all or parts of the AutoCAD drawing in many other formats besides postscript. Of the other formats, "wmf" and "bmp" are useful for use with other windows aplications while "dwf" is the drawing web format. DWF is for including drawings in a form viewable using a Web browser with a suitable plug-in.

Plotting multiple viewports in paper space

A simple title block for EXP-GIZA

Before embarking on the exercise on paper space and the MVIEW command, let's create a title block and margin in preparation for the final plot. This small drawing will be incorporated in the final plot of the pyramid using AutoCAD's XREF facility. In this short section we will make a new drawing of the margins and a title block for a standard A3 plot. Start AutoCAD if it is not still active and create a new drawing, picking **File/New**, and pick **Start from scratch** with metric units. It will be saved with the name **Title-A3**.

As with any other drawing, the first steps are to set up the limits and create any layers that will be needed. The limits for my A3 plotter are (0,0) to (420,297). Some plotters cannot plot up to the edge of the page and so you might have to adjust the limits accordingly.

Command: **LIMITS**
Reset Model space limits:

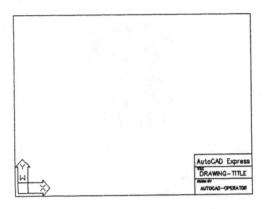

Figure 9.12 Margins and title box

ON/OFF/<Lower left corner> <0.00,0.00>: **<ENTER>**
Upper right corner <default>: **420,297**
Command: **ZOOM**
All/.../Window/<Realtime>: **A**
Command: **−LAYER**
?/Make/Set/...: **M**
New current layer <0>: **MARGIN**
?/Make/Set/...: **<ENTER>**

Now we can proceed with drawing the margins and title block shown in Figure 9.12. The title block includes some text.

Command: **RECTANG**
Chamfer/.../<First corner>: **5,5**
Other corner: **@410,290**
Command: **LINE**
From point: **315,5**
To point: **@65<90**
To point: **@100,0**
To point: **<ENTER>**

Now add the internal lines of the title block.

Command: **<ENTER>**
LINE From point: **315,50**
To point: **@100,0**
To point: **<ENTER>**
Command: **<ENTER>**
LINE From point: **315,30**

To point: **@100,0**
To point: **<ENTER>**

The next task is to add in the text and attributes. First make sure that the text font is SIMPLEX. Other fonts may not fit neatly in boxes with the text heights given below.

Command: **STYLE**

Change the Style name from "Standard" to **Simple** by picking the **New** button. Then pick **simplex.shx** from the Font Name list. Then pick **Apply** followed by **Close**. Command: **TEXT**

Justify/Style/<Start point>: **320,45**
Height <>: **3**
Rotation angle <0.00>: **<ENTER>**
Text: **TITLE**
Command: **<ENTER>**
TEXT Justify/Style/<Start point>: **320,25**
Height <3.00>: **<ENTER>**
Rotation angle <0.00>: **<ENTER>**
Text: **DRAWN BY**
Command: **<ENTER>**
TEXT Justify/Style/<Start point>: **C**
Center point: **365,55**
Height <3.00>: **7**
Rotation angle <0.00>: **<ENTER>**
Text: **AutoCAD Express**

The attributes will be used to input the drawing title and CAD operator's name. Each of the attributes must be visible and positioned using centered text. Define the drawing title first.

Command: **ATTDEF** or pick **Draw/Block/Define Attributes**
Attribute modes – Invisible:N Constant:N Verify:N Preset:N
Enter (ICVP) to change, RETURN when done: **<ENTER>**
Attribute tag: **DRAWING-TITLE**
Attribute prompt: **Enter drawing title:**
Default attribute value: **No title**
Justify/Style/<Start point>: **C**
Center point: **365,35**
Height <7.00>: **7**
Rotation angle <0.00>: **<ENTER>**

Now define the AutoCAD operator.

Command: **ATTDEF**
Attribute modes – Invisible:N Constant:N Verify:N Preset:N
Enter (ICVP) to change, RETURN when done: **<ENTER>**
Attribute tag: **AutoCAD-Operator**
Attribute prompt: **Enter your name:**
Default attribute value: **Anonymous**
Justify/Style/<Start point>: **C**
Center point: **365,12.5**
Height <7.00>: **5**
Rotation angle <0.00>: **<ENTER>**

These two attributes are now grouped to form a block. Whenever the block is inserted, you will be prompted to input the drawing title and your name.

Command: **BLOCK**
Block name (or ?): **TTEXT**
Insertion base point: **0,0**
Select objects: **365,35** 1 found
Select object: **365,12.5** 1 found
Select object: **<ENTER>**
Command: **QSAVE**

Now save the drawing by picking **File/Save** and give it the name **Title-A3**. The drawing is now ready to be used to generate plots. When the drawing is saved, it can be incorporated into any other AutoCAD file by either INSERT-ing it as a block or referencing it with the XREF command. You could also save it as a template file if you wished.

Paper space

You have already had a brief introduction to AutoCAD's concept of paper space in the last chapter. One of the significant features of paper space is the ability to output a number of different views of an object on one print out. If you print from model space you only get the active viewport.

By default, drawings are produced and viewed in "model space". That is, all the coordinates have been input relative to the world in which the object was drawn. When it comes to plotting, you could choose an origin position on the paper and also a rotation angle, hidden line removal, scale etc.

AutoCAD allows you to create a virtual page for plotted output. You do this by switching to "paper space". In paper space you can assign a page size, A3 for example, and a number of viewports. Different views of the object, say plan, elevation and perspective can be set up in each viewport and appropriate scales can be assigned. When all the views are correct the paper space can be printed at a scale of 1:1.

In EXP-GIZA, the paper space was then filled with three viewports as shown in Figure 9.13. Note, a margin of 10mm was included when positioning these viewports. These viewports, created by the MVIEW command are themselves AutoCAD entities. Thus, they belong to specific layers etc. They can be moved, copied and erased like other entities. Furthermore, they can overlap and even control the visibility of layers and hidden lines within any view. In Chapter 8 the three viewports were created on layer 0.

Remember, in order to use AutoCAD's paper space you have to set the value of the **TILEMODE** variable to 0.

Manipulating metaviews

In this section you will generate a standard engineering type drawing of the main pyramid constructed in Chapter 8. If you haven't got a copy of the drawing file, EXP-GIZA, you could improvise with any 3D object or just draw the outline of the pyramid as explained in Chapter 8. Open EXP- GIZA.dwg by picking **File/Open**. If you haven't already saved the TITLE-A3 drawing do so now. In order not to mess up the old Giza drawing pick **File/Save As** and give the new drawing name as **GIZAPLOT**. This procedure ensures that if anything goes wrong with the current exercise then at least we still have the original drawing.

The display will first resemble the view in Figure 8.25. To get to the situation shown in Figure 9.13 just set TILEMODE to 0. Pick **View/Tilemode** to do this. Then switch to Paper space by double clicking the **Model** button on the status bar, or type the command, **PSPACE**. The two smaller pyramids are still there – they are just out of view. The two toolbars in Figure 9.13 might come in handy. They can be found by picking **View/Toolbars** followed by checking **Viewpoint** and **Reference** from the list.

Command: **TILEMODE**
New value for TILEMODE <1>: **0**
Command: **PSPACE**

In Chapter 8 when you entered paper space for the first time in a drawing you had to set the paper limits. In this case the limits were set to A3 dimensions (420mmx297mm). For the plotting exercise you need to create a new layer for the viewport objects i.e. the frames of the viewports.

Command: **-LAYER** or use the Layer control button
?/Make/Set/...: **M**
New current layer <WALLS>: **PSVP**

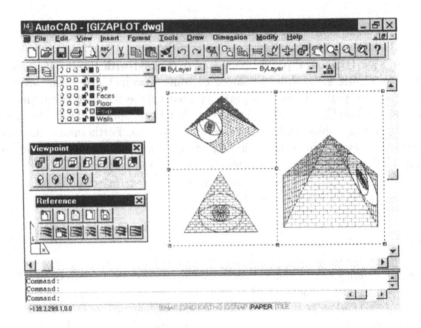

Figure 9.13 Starting Gizaplot

?/Make/Set/...: **C**

Color: **CYAN**

Layer name(s) for color 4 (cyan) <PSVP>: **<ENTER>**

?/Make/Set/...: **<ENTER>**

The LAYER command's "Make" option both creates a new layer and makes it the current one. The name PSVP is just my shorthand for "paper space viewports". Giving the layer a color cyan helps to distinguish the viewports from other objects on other layers.

The manipulation of the metaviews frames involves four key stages. Firstly, all the viewports will be changed to the PSVP layer. Then the large viewport on the right will be erased to be replaced by two smaller ones. The first will be obtained by copying one of the existing ones, while MVIEW will be used for the second. Once the viewports have been created the appropriate view points and scales for the plan, front elevation and side elevation must be chosen. The final, stage is the fine tuning of the viewports to get the plan and elevation views aligned.

Now execute the changing of layers. The easiest way to do this is to pick each of the viewport frames so that they become ghosted like in Figure 9.13.

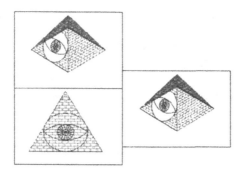

Figure 9.14 Copying a viewport

Then pick **PSVP** from the Layer pull-down list. The viewports should become cyan but remain ghosted. Press **Esc** twice to deselect the viewports.

Now erase the right hand viewport entity.

Command: **ERASE**
Select objects: **410,150**   (Point on right hand edge)
1 found
Select objects: **<ENTER>**

To fill the space left by this deletion, first copy the top left viewport from A to B as shown in Figure 9.14.

Command: **COPY**
Select objects: **10,200**   (Any point on the top frame will do)
1 found
Select objects: **<ENTER>**
<Basepoint or displacement>/Multiple: **10,285**   (A)
Second point of displacement: **210,180**   (B)

Everything about the viewport is copied including the VPOINT of the pyramid. This newest viewport overlaps the two others. In earlier versions of AutoCAD and when TILEMODE was 1 ie ON viewports could not overlap but could be arranged side by side like tiles – hence the term *tilemode*.

The final metaview shown in Figure 9.15 could also be created by copying, since it is the same size as the others. However, as a refresher on the MVIEW command we will create it from scratch. Pick **View/Floating Viewports** followed by **1 Viewport**.

Command: **MVIEW**
ON/OFF/Hideplot/Fit/2/3/4/Restore/<First point>: **210,147.5**

Other corner: **410,285**

Initially, the new viewport will display the same image as the currently active viewport in MODEL SPACE. The next stage involves going to model space, making each viewport in turn active and choosing the viewpoint and scale of the view. At the end of this stage you screen should have four viewports, as in Figure 9.15.

Command: **MSPACE** or double click the PAPER button

The upper right viewport should be outlined to indicate that it is the active one. When you move the cursor the cross hairs should only appear in this viewport. Elsewhere on the screen it appears as an arrow. If any other viewport is the active one, move the arrow cursor to the top left viewport and press the pick button.

Change the VPOINT to generate a side elevation. Pick **Left view** button from the Viewpoint toolbar. Then use the Zoom, Scale-XP option to set a 1:10 scale. The XP stands for "times paper space".

Command: **VPOINT**
Rotate/<View point> <current>: **-1,0,0**
Regenerating drawing.
Command: **ZOOM**
All/.../Window/<Realtime>:**.1xp**

This view should now look like that in top right of Figure 9.15. To get the other views as shown do the following. Make the upper left viewport the active one. Move the cursor into the upper left viewport and click the cursor. It should change from an arrow shape to the usual cross hairs. This view will now be changed to show the front elevation of the pyramid by picking the **Front View** button from the Viewpoint toolbar

Command: **VPOINT**
Rotate/<View point> <1.0,-1.0,1.0>: **0,-1,0**
Regenerating drawing.
Command: **ZOOM**
All/.../Window/<Realtime>:**.1xp**

Having got things down to size, the next thing is to make the all seeing eye visible again. The reason it is not clear at the moment is because of all the bricks belonging to the back wall. To make the bricks disappear as shown in Figure 9.15 we can freeze the WALLS layer for this viewport. Since the two small pyramids were created on the walls layer, they should be moved onto the Faces layer. All this can be done using the Layer pull- down list as before.

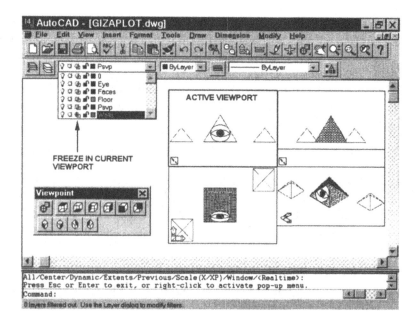

Figure 9.15 The right viewpoint

Pick the two small pyramids, then pick **FACES** from the Layer pull-down list. They should change to the red of that layer. Press **Esc** twice to deselect the pyramids. Now, go back the the Layer list and pick the icon looking like a sun superimosed on a rectangle (Freeze Thaw in Current Viewport) on the bottom line. This freezes Walls in the top left viewport.

Now make lower left viewport active and execute the PLAN command.

Command: **PLAN**
<Current>/Ucs/World: **W**
Regenerating drawing.
Command: **ZOOM**
All/.../Window/<Realtime>:**.1xp**

Finally, make the lower right viewport active and use either Zoom extents or Zoom Realtime to see the three pyramids. The exact magnification is not important for the isometric view.

This leaves the display as shown in Figure 9.15. The job is not quite finished yet, though. The three orthogonal views are slightly out of line with each other. In paper space you should be able to drop a vertical line from the front elevation to the plan and the side elevation should be on the same level as the front.

The first part of the fine tuning stage is done in paper space. You need to move the bottom left viewport a shade to the right. You also need to move the top right viewport slightly down. Then you will set up the hidden line removal for plotting the isometric view.

Switching is done with the **PSPACE** command. The shifting of the viewports is done with AutoCAD's standard **MOVE** command. Hidden line removal is one of **MVIEW**'s options.

Command: **PSPACE** or pick the Model/Paper button on the tool bar.

To align the plan view with the front elevation you need to move the lower left viewport to the right. The magnitude of the displacement can be got by using object snap to pick up the points A and B in Figure 9.16. The coordinate filters, .x, and .yz, are also used. Zoom in for a closer look if you need to be convinced that the views are misaligned.

Command: **MOVE**
Select objects: **100,10**   (Point on edge of bottom left viewport)
1 found
Select objects: <**ENTER**>
Base point or displacement: **intersec** of Pick point A
Second point of displacement: **.x** of **intersec** of Pick point B
(need YZ) **intersec** of Pick point A again.
Command: **MOVE**
Select objects: **315,285**   (Point on edge of top right viewport)
1 found
Select objects: <**ENTER**>
Base point or displacement: **intersec** of Pick point C
Second point of displacement: **.y** of **intersec** of Pick point B
(need XZ) **intersec** of Pick point C again.

This viewport should now appear shifted to the right as shown in Figure 9.16. A point to note about the previous manipulation is that you were able to snap to objects that have been defined in model space even though the paper space was active. However, it is not possible to make any physical alterations to the pyramid when paper space is active.

If you set "Hideplot" to "on" then a hidden line removal will be done in this view when the drawing is plotted. This is one of the options of the **MVIEW** command. You select a particular viewport by picking one of its edges.

Command: **MVIEW**
ON/OFF/Hideplot/Fit/2/3/4/Restore/<First point>: **H**
ON/OFF: **ON**

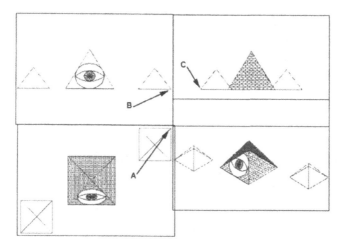

Figure 9.16 Fine tuning

Select objects: **410,100**   (Point on edge of bottom right viewport)
1 found
Select objects: **<ENTER>**

Cross referencing drawing files

Now that all the views have been set up we can add in the title block and
margin, Title-a3.dwg. This will be done using the XREF command. We could
use the INSERT command as an alternative.

The main advantage of using **XREF** over **INSERT** is that the objects of
the referenced drawing are not copied into the current one. They are simply
displayed at the same time. This helps to keep down the size of the current
drawing file. A second advantage is that any modifications made to the referenced drawing will automatically be made to the referring one.

While still in paper space we will **XREF** the drawing TITLE-A3. As **XREF**
will be executed while in paper space, the margins and title box also become
part of the paper space. It will also be attached to the current layer. Before
generating the final plot, we must make the edges of the viewports invisible.
To do this you must make the PSVP layer invisible. Change to layer 0 and
freeze PSVP.

Command: **−LAYER**
?/Make/Set/...: **S**
New current layer <PSVP>: **0**

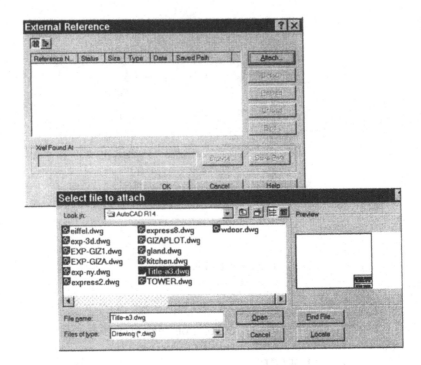

Figure 9.17 External Reference

?/Make/Set/.../Freeze/Thaw: **F**

Layer name(s) to freeze: **PSVP**

?/Make/Set/...: **<ENTER>**

You can now perform the XREF. Pick **Insert/Exteral Reference...** from the menu bar. This brings the External Reference dialog box shown in Figure 9.17. Pick **Attach** and select **Title-a3.dwg** from the list of files. Then pick **Open**. The Attach Xref dialog (Figure 9.18) pops up. Pick **OK** to attach the file with scales of 1 for x, y, and z.

Command: _xref

Attach Xref TITLE-A3: TITLE-A3.DWG Title-a3.dwg

TITLE-A3 loaded.

Insertion point: **0,0**

X scale factor <1>/Corner/XYZ: **<ENTER>**

Y scale factor (default=X): **<ENTER>**

Rotation angle <0.00>: **<ENTER>**

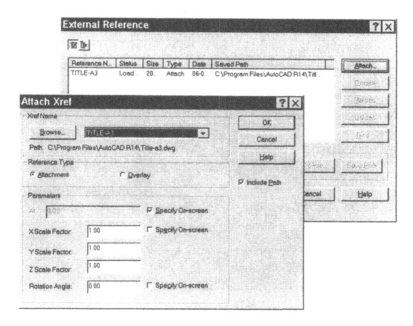

Figure 9.18 Attach Xref

Of the other options in the XREF command line "detach" allows you to remove a cross reference that was previously attached. The "bind" option will convert a XREF drawing into a conventional AutoCAD block. When you open any AutoCAD file the editor it automatically reloads any XREF's. The "Reload" option does the same within an editing session. This might be useful if a group of people are working in a series of interconnected drawings.

Dependent blocks

If you look at the list of blocks you will find that TITLE-A3 is defined as an external reference block. The another block TITLE-A3|TTEXT is also defined as a dependent block. This is, of course the TTEXT block that was defined as part of the TITLE-A3 drawing earlier. Even though the block appears on the list it is not accessible in its current form.

Command: **BLOCK**
Block name (or ?): **?**
Defined blocks.
   TITLE-A3   Xref: resolved
   TITLE-A3|TTEXT   Xdep: TITLE-A3

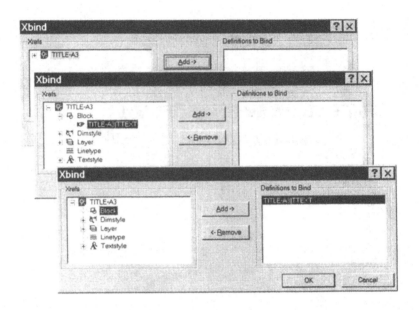

Figure 9.19 Xbind

User blocks	External References	Dependent Blocks	Unnamed Blocks
0	1	1	4

In order to be able to use the TTEXT block it has to be bound to the current drawing. This can be done by binding the whole of TITLE-A3 which converts it to a conventional block and also converts the nested block. It can also be done by just binding the TTEXT block and leaving the rest of TITLE-A3 as an XREF. A special command, **XBIND** does the latter.

Command: **XBIND**

The Xbind dialog box appears as shown in Figure 9.19. Initially, this just shows Title-a3 in the Xrefs section. Double click on the "TITLE-A3" text and then double click on the "Block" to see the dependent block name. Finally, pick the TTEXT and the **Add** button followed by **OK**.

At this point the block will become a conventional part of the current drawing. Its name will change to "TITLE-A3$0$TTEXT". The | is changed to 0. This new block can now be inserted. You may have to input the title and name using the dialog box.

Command: **INSERT**
Block name (or ?): **TITLE-A3$0$TTEXT**

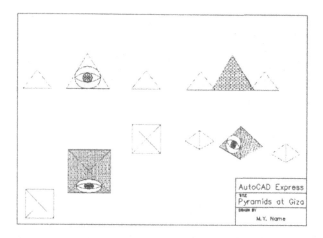

Figure 9.20 The final plot of Giza

Insertion point: **0,0**
X scale factor <1>/Corner/XYZ: **<ENTER>**
Y scale factor (default=/x): **<ENTER>**
Rotation angle <0.00>: **<ENTER>**
Enter the drawing title: <No title>: **Pyramids at Giza**
Enter your name: <Anonymous>: **M.Y. Name**
Command: **QSAVE**

Save the drawing and your screen should now look like the plot shown in Figure 9.20 without the hidden line removal. You can plot it now to see the hidden line removal take place in the isometric view.

A note on XREF files

The XREF'fed file cannot be altered from inside the referring file. Thus you cannot change the contents of the drawing TITLE-A3 from inside the drawing "GIZAPLOT". If you write your name within the title box, then that text will be part of the current drawing only. However, you can use object snap to points in the XREF'fed drawing.

The referenced drawing will be associated with whatever is the current layer when the XREF command is issued. Above, the title box and margins are attached to layer 0. However, it is not as simple as all that. A new layer, called "TITLE-A3|MARGIN" will also be created automatically. The visibility of the margins etc is controlled by both layer 0 and TITLE- A3|MARGIN. If either is frozen then the margins will be invisible.

When you exit a drawing that contains XREFs the information from the referenced file is deleted from the current one. Only the actual reference is retained. This is used the next time you edit the drawing to resolve the XREF and display the drawing in full.

You can only xref a drawing that has been created in model space. Paper space entities are not copied.

The softcopy

Safety first

In any drawing office, security and archiving are given great priority. This must be extended when drawings are stored on magnetic media. Some users will have access to sophisticated CAD management programs that take care of this. Many will just be relying on sensible practices.

The first sensible practice is to save the drawing regularly. In particular the drawing should be saved before "hazardous" operations such as plotting or using HATCH. AutoCAD provides one backup copy in the same directory as the drawing. You, the sensible user, should have another backup on another disk. A further safe copy should be stored in a separate and safe location. If one copy gets corrupted make sure it is replaced immediately, don't put it off.

Large volume archiving is best achieved by copying to a tape storage device or laser disk. This normally requires specific hardware attached to your computer.

If you are archiving drawings that contain xref's then it is recommended that you bind the referenced drawings to the archived one using XBIND. Alternatively, you can archive all the xref's as separate files along with the drawing that uses them. You are also recommended to bind all references when creating DXF files for interchange with other CAD systems.

Copying to other Windows applications

A great advantage of AutoCAD running under Windows is that it opens up all the possibilities of that environment. Not only can you resize the drawing window and simultaneously run other applications but you can use Windows to communicate between AutoCAD and these applications. This means that you can copy text and drawing notes from your Windows wordprocessor to AutoCAD and copy drawings into your wordprocessor.

To copy text into an AutoCAD drawing you can use the Windows clipboard. When there is something in the clipboard, AutoCAD's **Edit** pull-down menu allows you to **Paste Command**. This should only be picked when the

MTEXT or DTEXT command is running and prompting for text. MTEXT itself has an text IMPORT button which allows you to bring a basic text file or a file in Rich Text Format (rtf). Applications such as MS Word can produce rtf files that retain all the formatting and font information.

This text then becomes a normal AutoCAD text entity. Note that there is no active link between the original wordprocessor file and the drawing. The clipboard merely did a copy operation. If you change the text in the original file it will have no new effect on the AutoCAD text.

It is possible to copy an AutoCAD object to any other windows application via the clipboard. This can be done using **Edit/Copy** which runs **Copyclip** for copying individual objects. these can then be pasted into other applications (or AutoCAD files). They retain their vector information.

The **Edit/Copy Link** command differs from Copyclip in that a link is set up between the receiving application's file and the AutoCAD file. IF the AutoCAD file is subsequently updated, the other file can also be updated.

Similar to the copying from AutoCAD, one can paste vector and bitmap information into an AutoCAD drawing. Paste Clip and Paste Special are the corrseponding commands on the Edit menu.

Object linking and embedding

Many Windows applications support Object Linking and Embedding (OLE). This allows you to create active links between applications. AutoCAD supports OLE as a server application and as a client. To use OLE you need the other application to support it as a server or client. When AutoCAd is the server it means that AutoCAD is the source and the other application receives the drawing. For example, you can link or embed an AutoCAD drawing in a Microsoft Word document. When AutoCAD acts as a client it means that information from another such as a spreadsheet from MS Excel has been inserted in the drawing file.

The difference between OLE and the simple cut and paste done above is that when a drawing is embedded in a document it can be edited using AutoCAD or the server application. The difference between linking and embedding is subtle. When an AutoCAD link is established, any changes to the original drawing will be applied in the document by picking UPDATE from the Edit/OLE menu.

Hazard warning for OLE users. While OLE is an established mechanism for communication between computer applications, many personal and office work practices make the management of the links quite tricky. Unless a set of links is thought through and properly managed it will soon become unwieldy and maybe unworkable. Always treat OLE with caution and ask yourself if it is really necessary? While embedding objects loses the update facility it is generally more robust.

Summary

This chapter has covered the procedures for producing hardcopy output of AutoCAD drawings. It has also introduced a number of Windows functions for copying AutoCAD drawings to other applications.

You should now be able to

Set up the printer/plotter parameters.
Plot multi-colored drawings on pen plotters.
Exporting postscript output to a file.
Use MVIEW to make overlapping viewports.
Make layers frozen in individual viewports.
Plot multiple viewports with hidden line removal.
Use cross referenced files.
XBind dependent blocks.
Copy to the Windows clipboard.

Appendix A CONFIGURATION

General

Configuring AutoCAD to work correctly used to be a major chore. Now, thanks to Windows NT it is a breeze. All the configuration settings are available through one dialog box (albeit one with many parts).

Start AutoCAD and pick **Tools/ Preferences** from the menu bar. The Preferences dialog box (Figure A.1) allows you to configure everything from the AutoCAD Graphics Window to printers and pointing devices. This short appendix covers a few useful topics from the Preferences dialog box plus a few neat and simple customizations.

AutoCAD Preferences

Profiles

Have you ever come to an AutoCAD session only to find that some other joker has completely reconfigured your system deleted all your toolbars and generally messed things up? With the Profiles in the Preferences dialog you can store as many different configurations for as many different users. You can Copy the default profile and rename it with a little extra description. If you are paranoid about others corrupting your profile you can also export it to an external file. The exporting should be done after all the configuration changes have been implemented.

Display

Figure A.2 shows the Preferences dialog when the **Display** tab is picked. Picking the **Colors** button allows you to reconfigure the way the AutoCAD screen is displayed. In particular, this is useful for setting whether the AutoCAD screen is white with dark vectors or black with light vectors.

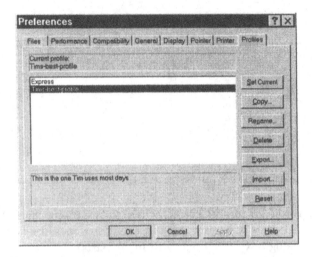

Figure A.1 Preferences dialog box

Project file path

The Files tab in Figure A.3 allows you to set the folders for various types of file. A useful facility here is to specify the search path associated with a particular project. This can be linked to the current drawing via a system variable as indicated on the figure.

There are many other things that can be set up via the Preferences dialog. It is worth exploring the other tabs via the Help button. Most of the items are self explanatory. The most useful items have been described above and on page 33 where automatic saving was discussed.

Customizing a toolbar

One of the key features in making AutoCAD the most successful CAD program ever is its customizability. A whole industry of 3rd party add-on software has grown up around AutoCAD as a result of its open architecture and accessibility. There are some things in AutoCAD which actively promote users to create their own macros and user defined features. Making your own toolbar is one of the easiest to implement. It is also one of the best productivity tools you can have.

AutoCAD is a BIG program with hundreds of toolbar buttons. These are generally arranged in themes e.g. Drawing commands, Modifying commands etc. Here is how to make your own toolbar with a few much used buttons.

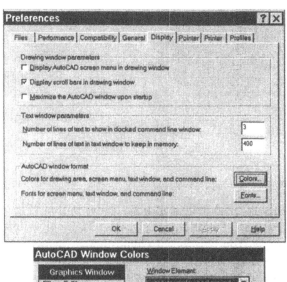

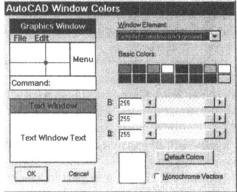

Figure A.2 Preferences – Display

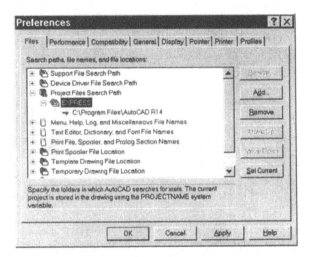

Figure A.3 Preferences – Files

Pick **View/Toolbars...** and then pick the **New** button (Figure A.4). Call the toolbar **EXPRESS TOOLS** and pick **OK**. The toolbar initialy appears as a small empty one. You may have to hunt around the sxcreen to find it.

Now, add a few buttons to the new toolbar. Pick the **Customize** button on the Toolbars dialog and from the Categories list of the Customize Toolbars dialog select **Draw**. Note that there are a lot more icons here than on the Draw toolbar. With the mouse pick one of the icons and drag it to the Express Tools toolbar and just drop it in. In the figure I have included 3D Polyline, Polygon, Divide and Donut on the top row. These were all selected from the Draw category. The 4 buttons on the bottom were taken from the Modify category. Can you figure out what commands they represent?

If you want to remove a button from a toolbar, simply drag it from the toolbar and drop it in the drawing area of the screen. You can alter the properties of a button by pressing the right mouse button to select it. In the button properties dialog box you can edit the name of the command, the macro and even edit the icon itself.

With the idea that you can now create your own AutoCAD commands and icons the AutoCAD Express comes to a close. This is far from the end, however. Far from it indeed – this is the start of something big.

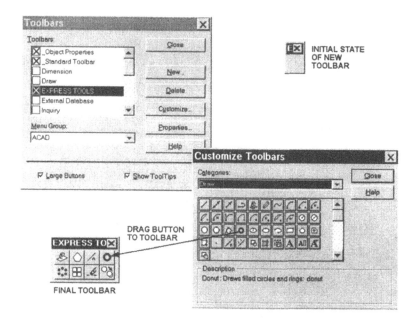

Figure A.4 Customizing a toolbar

SUBJECT INDEX